Food System
Transparency

Advances in Agroecology

Series Editors

Stephen R. Gliessman, University of California, Santa Cruz, California

Helda Morales, Agroecology Group, El Colegio de la Frontera Sur, San Cristóbal de Las Casas, Mexico

Environmental Resilience and Food Law
Agrobiodiversity and Agroecology,
edited by Gabriela Steier, Alberto Giulio Cianci

Political Agroecology
Advancing the Transition to Sustainable Food Systems,
authored by Manuel González de Molina, Paulo F. Petersen, Francisco Garrido Peña, Francisco R. Caporal

Urban Agroecology
Interdisciplinary Research and Future Directions,
edited by Monika Egerer, Hamutahl Cohen

Subtle Agroecologies
Farming with the Hidden Half of Nature,
edited by Julia Wright, Nicholas Parrott

Food System Transparency
Law, Science and Policy of Food and Agriculture,
edited by Gabriela Steier, Adam Friedlander

For more information about this series, please visit: https://www.crcpress.com/Advances-in-Agroecology/book-series/CRCADVAGROECO

Food System Transparency

Law, Science and Policy of Food and Agriculture

Edited by
Gabriela Steier and Adam Friedlander

CRC Press
Taylor & Francis Group
Boca Raton London New York

CRC Press is an imprint of the
Taylor & Francis Group, an **informa** business

Front cover photograph by Adam Friedlander
Back cover photograph by Paul De Filippi

CRC Press
Boca Raton and London

First edition published 2021

by CRC Press
6000 Broken Sound Parkway NW, Suite 300, Boca Raton, FL 33487-2742

and by CRC Press
2 Park Square, Milton Park, Abingdon, Oxon, OX14 4RN

ISBN: 978-0-367-77412-7 (hbk)
ISBN: 978-0-367-44036-7 (pbk)
ISBN: 978-1-003-00755-5 (ebk)

Typeset in Palatino LT Std
by KnowledgeWorks Global Ltd.

Dedication

My parents
My grandparents
My sons
My husband

— Gabriela Steier

Mom, Dad, Benjamin, Jasmine,
my late grandparents,
and
all of my mentors, friends, and family.

— Adam Friedlander

Contents

Part I: The big picture

Part II: Food safety and health

Preface

> In a time of turbulence and change, it is more true
> than ever that knowledge is power; for only by true
> understanding and steadfast judgment are we able
> to master the challenge of history.
>
> **John F. Kennedy, 1962**

This volume has been written and finalized during the first six months of the COVID-19 pandemic. The editors are considered Millennials and have lived through these turbulent times with open eyes, minds, and hearts. As such, the goals for this book are founded most deeply in the effort to ascertain the power of information within the food system.

There are few powers greater than information and knowing better means one can do better. Specifically, if one understands where the information about one's food lies or how it can be obtained, such knowledge results in unbelievable potential to prevent outbreaks, improve food quality and even safety, and to prevent food insecurity. Only if the food system is transparent, can actors and stakeholders, regulators, and consumers tug at the right end of this vast network of convoluted and intertwined information. Hopefully, information about our food system will reach consumers – and resonate with their conscience and intellects. It is with this in mind, that this volume has been compiled and edited.

Gabriela Steier
Boston, Massachusetts

An international network of scientific and legal scholars deliver timely masterpieces to enhance the public's understanding of food system transparency. But what is transparency and how does the food system succeed alongside societal forces? While the definition of transparency may evoke varying opinions and emotions, there are undeniable and urgent opportunities to help make the food system work for all. Throughout our lifetime, we must keep eternal optimism that our collective and goodwilled efforts can eradicate malnutrition, poverty, health disparities, racial and ethnic injustices, and socioeconomic hardships – particularly among civilization's most vulnerable and neglected communities.

Food System Transparency: Law, Science and Policy of Food and Agriculture was written, edited, and published in the midst of COVID-19, which intensified geographic, political, economic, scientific, racial, and other social inequities within the global food system. But these weaknesses present crucial opportunities for progress and justice beyond the pandemic. This volume was edited to inspire all participants within the food system to get involved in the agricultural process while considering how our daily actions impact others within the diverse, complex, and essential food system.

Adam Friedlander
Washington, DC

Introduction

This book is structured in three parts:

    I. The Big Picture
    II. Food Safety and Health
    III. The Global View

Each part adds to the whole but zooms in through a lens that is unique to that chapter alone. As such, students, practitioners, regulators, policy-makers, industry stakeholders, and anyone who wishes to gain a deep understanding of the intricate working of food system transparency will find satisfaction in this volume. The theme of transparency is interwoven into discussions about the importance of agricultural information sharing, and how complicated policies and circumstances can make information inaccessible. While the chapters can be read individually, it is beneficial to see the context that they each provide to holistically grasp fundamental and diverse concepts of food system transparency.

In Part I, the contributing authors provide a host of information on food safety modernization, food fraud, food defense, agroecology, and ag-gag. The large-scale international regulation through the Codex Alimentarius is outlined and the greater context with the sustainable development goals of the United Nations are included.

Part II zooms in on food safety and health within North America. Practical case studies explore foodborne outbreak investigations, COVID-19 consequences and supply chain innovation for emerging plant-based crops.

Finally, Part III, introduces specialized aspects of food system trans-parency in major trading areas. The European Union, China, and Africa share many opportunities and challenges (technology, collaboration, reg-ulation, etc.), but each are distinguished through varying approaches to solve problems according to law.

To facilitate further context and materials within each chapter for classroom use, the editors' provide commentary that blends legal and sci-entific expertise.

Acknowledgments

The editors thank all of the contributors of this volume. It is only with their collaboration, expertise, and dedication that advances in the field are possible. It has been an honor and a privilege to be able to brainstorm, draft, and ultimately publish this quintessential volume on a topic that is dear and important to all of us.

Special gratitude goes to our families, who patiently blocked out time for us to discuss, manage, edit, and write until this volume was complete. Without their teachings, patience, support, and love, none of this work could have been possible.

Deep gratitude goes to Alice Oven and Damanpreet Kaur from Taylor & Francis/CRC Press whose unwavering support has become a standing leg for this book. Special thanks are also extended to Balraj Mani, senior project manager at KnowledgeWorks Global Ltd., for all the edits and patience with our manuscript.

Our readers deserve heartfelt thanks for propelling our scholarship forward in study, discussion, criticism, and classroom use.

Editors

Gabriela Steier is a lawyer, educator, speaker, and scholar focusing on food law and regulation. She has established herself as an internationally recognized expert in her field. Her areas of focus are food system regulation in the EU and U.S., environmental and climate change law and policy. She is a Part-Time Lecturer at Northeastern University, where she teaches several courses on food regulation, food safety modernization, and international food trade. She also teaches as an Adjunct Professor at the Duquesne University School of Law in Pittsburgh, and as a Visiting Professor at the University of Perugia, Italy. As Founder of FoodLawInternational.com, she manages a vast network, hosts lecture series, and promotes academic scholarship in her field. She has published widely in both textbooks and peer-reviewed journals. Dr. Steier has a BA from Tufts University, a JD from the Duquesne University School of Law, an LLM from the Vermont Law School, and a Doctorate in Comparative Law from the University of Cologne, Germany. She is based in Boston, Massachusetts.

Adam Friedlander is a food scientist focused on modernizing food safety approaches within industry, academic, and consumer communities. He is a Manager, Food Safety and Technical Services at FMI – The Food Industry Association. As a dynamic member of the Institute of Food Technologists (IFT), International Association for Food Protection (IAFP), Conference for Food Protection (CFP), and Association of Food and Drug Officials (AFDO), Friedlander collaborates with food industry experts to help keep the global food supply safe. He serves as an Ambassador with the Partnership for Food Safety Education (PFSE) to equip consumers with foundational food safety knowledge that helps reduce foodborne illness risk. Friedlander graduated from Cornell University with a Bachelor of Science in Food Science and Operations Management, and a minor for Music. He received a Master of Science in Regulatory Affairs of Food and Food Industries from Northeastern University.

Junior Editor

Christian Mulgrew de Laire is a graduate from The College of the Holy Cross where he studied Economics and French. His concentration consisted of anti-trust, sports law, and case studies. While there he also studied abroad in France at the prestigious Science Po Strasbourg, taking classes in European Economics and Law. de Laire is fluent in English and French. As junior editor of this volume, he has edited several chapters and created figures. He is an aspiring law student and has great academic potential.

Contributors

Rosemary Agbor is an independent educational consultant and social reformist interested in the intersectionality of the law and social policy pertaining to gender and sustainable development, education, economic inequality and crime. She holds an LLB, LLM, MS in Criminal Justice and an EdD in Education from Grambling State University. She has previously taught undergraduate and graduate level courses in Education and Criminology at Grambling State University. She also sits on the advisory board of a U.S. charitable organization and a West African based NGO committed to promoting sustainable development.

Nicole Arnold is an Assistant Professor in Nutrition Science at East Carolina University. Her research aims to assess the current landscape of various food-based activities in practice, in order to make recommendations for more effective food policies and procedures. Her research interests include food safety education and interventions, risk communication, consumer perceptions of food and processing methods, and food labeling. She is a member of the International Association for Food Protection and serves as a Councilperson for the Food & Laws Council of the Conference for Food Protection. Arnold graduated from North Carolina State University with a Bachelor of Science and Master of Science in Food Science. She obtained her PhD in Food Science and Graduate Teaching Certificate from Virginia Tech.

Stephanie Brown is a Food Safety Specialist at the Oregon State University's Food Innovation Center in Portland, Oregon. As a member of the Kovacevic Lab, Brown works on projects related to the diverse research interests of the lab including antimicrobial resistance, stress response, and pathogen survival. She also works as the coordinator for the Western Regional Center to Enhance Food Safety. Before coming to Oregon State University, Brown completed her MS degrees in Animal Science from the University of Connecticut (2018) and Food Science from the University of Georgia (2015). She also completed her BS degree in Food

Science and Technology from the University of Georgia in 2013. She is an active member of the International Association for Food Protection, Institute for Food Technologists, American Dairy Science Association, and Phi Tau Sigma Honor Society. In her spare time, Brown is completing her doctoral dissertation on antimicrobial strategies to control *Listeria monocytogenes* in high moisture cheeses and the impact of these strategies on *L. monocytogenes* virulence.

Darin Detwiler is an internationally recognized and respected food policy expert with over twenty-five years of experience in shaping federal food policy, consulting with corporations, contributing thought leadership to industry events and publications, and advising industry, NGOs, and government agencies on food safety and authenticity issues in the United States and abroad. In 2018, Detwiler received the International Association for Food Protection's Distinguished Service Award (sponsored by *Food Safety Magazine*). Dr. Detwiler is the Assistant Dean at Northeastern University's College of Professional Studies and the Lead Faculty of the Master of Science in Regulatory Affairs of Food and Food Industries, where he has specialized in food safety, global economics of food and agriculture, blockchain, and food authenticity. He is the author of *Food Safety: Past, Present, and Predictions* (Elsevier Academic Press, 2020) and the editor of *Building the Future of Food Safety Technology: Blockchain and Beyond* (Elsevier Academic Press, 2020).

Wele Elangwe is currently an Adjunct Lecturer of Aviation Law and Broadcast Law at University of Maryland Eastern Shore where she also serves as Director of Graduate Student Services. She has worked as foreign legal consultant in Germany (2017) and with the United Nations in Central Africa (2006–2008 and 2012). She holds an LLM in International law from Indiana University Robert H. McKinney School of Law and a Maîtrise en Droit in Business Law from the University of Yaoundé II, Cameroon. She has authored many publications on fundamental rights, agricultural law, food law and food safety transparency, labor law, and gender equality and leadership. She is an advocate for sustainable development and a certified forensic investigation professional (CFIP) with a penchant for justice, accountability, transparency, and fairness.

Justin Falardeau is a PhD candidate at the University of British Columbia under the supervision of Dr. Siyun Wang. His research focuses on the inhibition of foodborne pathogens by the cheese microbiome. During his studies, Falardeau has received several major awards including the Alexander Graham Bell Canada Graduate Scholarship and the Izaak Walton Killam Memorial Pre-Doctoral Fellowship. He is an active member of the International Association for Food Protection (IAFP) and has been the President and Vice President of the British Columbia Food Protection Association (BCFPA). Falardeau holds a BSc in Food Science and Nutrition from Carleton University, where he was awarded the Senate Medal for Outstanding Academic Achievement, and an MSc in Food Science from the University of British Columbia.

Paul De Filippi has performed in a regulatory, environmental health, and safety capacity for both governmental and private industry. Most recently within the diverse Hawaiian agricultural industry, he has worked as a regulatory compliance inspector for the State of Hawaii under the United States Environmental Protection Agency and, in parallel, as the sole Manager of a material start-up fruit farm, food production, and consultation operation. He serves on the Board of Directors for the Hawaii Master Food Preservers and the Hawaii Tropical Fruit Growers Associations. He received his Bachelor of Science degree in Environmental Science from Northeastern University and is currently working toward a Master of Science from Northeastern University in Regulatory Affairs of Food and Food Industries.

Karen Fong is a microbiologist interested in food safety and effective mitigation strategies to combat human pathogens in the food chain. Dr. Fong completed her PhD in Food Science and subsequent postdoctoral work at the University of British Columbia where she focused on investigating bacteriophage-based applications for biocontrol of *Salmonella* in the food industry. She is currently a research scientist at Agriculture and Agri-Food Canada (AAFC) in Summerland, British Columbia, Canada.

Sam Jennings has been working for Berry Ottaway & Associates Ltd. since 2001. She provides advice to industry and governments globally on scientific, technical, and regulatory aspects of food. She offers support for dossier production for submission to the Commission and EFSA, particularly for novel foods, food additives, and health claims, and is currently an industry adviser on the permitted use of nutrition and health claims under the EU Regulation. Jennings is a Fellow of the Institute of Food Science and Technology (IFST) in the UK and Chair of the IFST Food Regulatory Steering Group. She is also a member of the Institute of Food Technologists in the United States, and of the Society of Dairy Technology in the UK. Sam is Vice-Chair of the UK Government's Office for Product Safety and Standards Business Expert (Food Standards & Labelling) Group and is a member of the UK Food Standards Agency's Working Group on Food Standards. Since 2006, she has been Technical Adviser to the Council for Responsible Nutrition UK (CRN UK). She has been a member of the International Alliance of Dietary/Food Supplement Associations (IADSA) International Technical Working Group since its conception in 2010 and was Chair of this group from 2012 to 2017. She was a member of the IADSA Scientific Council from 2006 until its cessation in 2018.

Autumn Johnson holds a JD from the University of Oregon, an MBA from Seattle University, and is a PhD candidate at Boise State University in the Public Policy and Administration program. She is an Energy Policy Analyst with Western Resource Advocates, a non-profit environmental organization focused on decarbonizing the electric grid. She also teaches agriculture and energy law at Concordia University School of Law.

Fabrice Mbala is a senior partner with Justice Law Firm, Cameroon. After earning his Bachelor of Laws (Hons) degree in 2002 from the University of Yaoundé II and a Maîtrise en Droit in Business Law from the University of Dschang (2004), Barrister Mbala was admitted to the Cameroon Bar in 2011. He has since gone on to fulfill with purpose his passion for the Law, taking up a variance of roles in corporate affairs, notably as Contracts Manager for West and Central Africa for oil and gas titan, Schlumberger. Currently, he is part of the legal team for Cameroon's national brewery corporation, Le Brasseries du Cameroun, where his advanced knowledge and expertise in food safety, corporate and regulatory affairs help ensure legal compliance at the brewery.

Alexey Petrenko is a regulatory consultant specializing in global food safety, nutrition policies, product standards, and market regulations. He is an advisor on Codex Alimentarius documents and procedures and a regular member of the Russian delegations in various committees. He is also an expert in food regulations of the Eurasian Economic Union and member of several working groups engaged in developing new regulations. Dr. Petrenko works with governments, industry, and NGO sectors across the world providing regulatory advice and training on practical aspects of global and regional food law. For many years, Dr. Petrenko was a member of the UK-based academic research group. He holds a PhD in organic chemistry and is the author of twenty-five scientific publications.

Juanjuan Sun is an Associate Professor at Hebei Agricultural University and Research Fellow of Center for Coordination and Innovation of Food Safety Governance at School of Law of Renmin University of China. Her research focuses on regulatory theory, food law, and agricultural law.

Victor Tutelyan is a globally renowned nutrition scientist having served as a head of Russia's Federal Center of Nutrition Food Safety and Biotechnology for over twenty years. Since 1993, he has been the national Codex Contact Point and a regular head of the country's delegations in the Codex Alimentarius Commission. He is also an expert on Food

Safety in the World Health Organization (WHO), and member of the Joint Committee of FAO/WHO on Food Additives and Contaminants. Professor Tutelyan is a full member and a Deputy General Secretary of the Russian Academy of Sciences, a full member of the National Academy of Sciences of Armenia and the Academy of Sciences of Ukraine. He is the author of more than 560 scientific works, 15 of which are monographs, about 300 are methodological and regulatory documents, and 49 are patents on inventions.

Siyun Wang is an Associate Professor and the principal investigator of the Wang Lab of Molecular Food Safety (http://foodsafety.landfood.ubc.ca) at the University of British Columbia (UBC), where she teaches and conducts research on reducing foodborne pathogens in the food supply system. Over the past ten years, she has authored over fifty peer-reviewed papers and book chapters on food safety. Dr. Wang received her BSc in Pharmacy from Fudan University in Shanghai and PhD in Biology from Illinois Institute of Technology in Chicago. Prior to joining UBC in 2013, she was a postdoctoral associate at Cornell University. Siyun currently serves as an Associate Editor of *Current Research in Food Science*, and an editorial member of *Journal of Food Protection and Food Microbiology*.

Lily Yang is an enthusiastic and collaborative food scientist and science communicator. She is a Manager, Food Safety at The Acheson Group. She specializes in the impacts of consumer and stakeholder attitudes, behaviors, and technological utilization within the food system. Her food safety approach is through scientific and cultural inclusivity, literacy, accessibility, and capacity-building to nurture trust and bridge knowledge gaps to "meet people where they are." She is an active member of the International Association for Food Protection (IAFP), Conference for Food Protection (CFP), Institute of Food Technologists (IFT), Phi Tau Sigma Honor Society, National Association for Media Literacy Education (NAMLE), and Society of Asian Scientists and Engineers (SASE). She serves on the editorial board for the *Journal of Food Protection*. Yang graduated from the University of California, Davis with a BS in Food Science and Technology and minors in Plant Biology and English. She has also previously worked at the USDA and the International Food Information Council. Later, she received her MS and PhD in Food Science and Technology from Virginia Tech.

Jennifer Williams Zwagerman is the Director of the Drake Agricultural Law Center and an Assistant Professor of Law at Drake University Law School. She is a 2004 graduate of Drake Law School, where she obtained her certificate in food and agricultural law and served as editor-in-chief of the *Drake Journal of Agricultural Law*. She also received her LLM in Food & Agricultural Law from the University of Arkansas School of

Law – Fayetteville. Prior to joining Drake Law School, she was an attorney in the Des Moines office of Faegre & Benson (n/k/a Faegre Drinker) with a national food and agribusiness practice and served as a law clerk to the Honorable David R. Hansen on the Eighth Circuit Court of Appeals. Zwagerman served as the 2017–2018 president of the American Agricultural Law Association, is the 2020–2021 Chair of the Board of Trustees for the Academy of Food Law & Policy and is a member of the Iowa State Bar Association Agricultural Law Council and Small/Solo Section Council.

part one

The big picture

chapter one

Food fraud and food defense: Food adulteration law and the sustainable development goals (SDGs)

Darin Detwiler

Contents

EDITORS' NOTE: THE IS-AND-IS-NOT OF FOOD SYSTEM TRANSPARENCY AND THE FDA'S FOOD SAFETY MODERNIZATION ACT (FSMA)

In the first chapter of this volume, Dr. Darin Detwiler introduces the fundamental concepts of food fraud and food defense to shed light on the collaborative work needed to continue removing bad-faith actors in the global supply chain and protect public health. Detwiler analyzes the historical, geographical, regulatory, political, legal, scientific, social, and economic factors associated with food fraud and defense to highlight how food system transparency has modernized within the last three hundred years.

While the first British food fraud laws were developed for tea and coffee, these laws influenced American history and eventual food safety expectations for global consumers. By exploring the role of United States Department of Agriculture (USDA) Chief Chemist Dr. Harvey Washington Wiley's "Poison Squad" experiments in the late 19th century and Upton Sinclair's novel "The Jungle" at the turn of the 20th century, Detwiler aims to magnify how America's first consumer protection laws pioneered current food safety expectations.

This chapter concludes by identifying recent foodborne outbreaks and the criminal and civil implications associated with these events. In an effort to reduce Norovirus, *Salmonella, Listeria, Escherichia coli O157:H7, Campylobacter,* and Hepatitis A infections, as well as protect consumers from severe allergic reactions, Dr. Detwiler introduces how the Food and Drug Administration (FDA) modernized the food regulatory system with the passage of the landmark Food Safety Modernization Act (FSMA) of 2011.

The following chapter provides a solid basis for all those seeking to explore food fraud, food authenticity, and food defense. Gaining this foundational understanding is at the backbone of comprehending what food system transparency means. The is-and-is-not of food system transparency begins with this following masterfully crafted chapter.

Introduction

Not even as far back as fifty years ago, American consumers still made food-buying decisions based on three primary questions – "Will it satisfy my desire for taste?" "Will it be enough to fill me up?" and "Can I afford it?" Today, consumers ask new questions in their quest for validation of their food source's reputation.

A number of forces changed the nation's food supply system over the past hundred years, or so. America had already shifted from a predominantly agrarian and rural society to an industrial one where the largest percentages of the population lived in denser urban centers. The arrival of immigrants at the turn of the 20th century not only diversified foods and flavors, but also played a significant role in the development of the produce industry on the West Coast. The Great Depression, the Dust Bowl, and two World Wars brought about rationing, new methods for preserving foods, and new policies regarding food sustainability. The invention of new technologies prompted radical changes in how we related to food in our homes and out in town. Foods could now be frozen for ease in mass distribution and retail of entire meals. In the 1950s and 1960s, the automobile brought about fast food, where people went to the restaurants and could eat inside or outside – still in their cars. Half a century later, cars (along with computers and cell phone apps) would bring food from the same restaurants and from grocery stores to people's front doors.

Perhaps these changes in technology reflect a change in consumer behavior. In 2015, the U.S. Department of Commerce released data that showed how, for the first time ever, the amount of money that American consumers spent at restaurants and bars overtook the amount spent at grocery stores (Jamrisko, 2015). Even more recent research highlights that Millennials (those people born between 1981 and 1996) tend to focus purchase decisions on convenience – such as ready-to-eat foods – rather than making food at home. This age group also tends to spend less money, overall, on food at home and makes fewer trips to the grocery store (Kuhns & Saksena, 2017).

With the myriad changes in food from the farm to the fork, and in how consumers' behavior around and related to food, opportunities for failure and for crime have not only increased, but have, unfortunately, become reality. Once taken for granted in terms of its safety, food has acquired a history of leaving far too many families to face the true burden of disease and of crime. Digging deeper into the history of policy for food fraud and adulteration can shed more light on the thoughts and language that shaped current policy.

Tea and coffee: A brief history of adulteration policy in Britain

Drinking tea became commonplace in London and spread throughout England by the late 17th century. Through the 18th century, the British replaced beer and ale with tea as their national drink. Developing a tea ceremony, the British established teatime in their culture, customs, and etiquette. Further, tea was viewed as part of an elevated diet for the social and wealthy. With this notion came the drive for fine ceramic and porcelain production, as well as for import of the pottery. Soon, after tea became hugely expensive and heavily taxed, British lawmakers took action to curb the growing fraud in tea and in coffee. The correlation between teas' evolution to a commodity food and its adulteration illustrates food system transparency issues and principles that permeate many other foods.

In 1718, for instance, Great Britain's Parliament passed the ADULTERATION OF COFFEE ACT of 1718 (5 Geo. 1 c. 11) making it illegal to debase coffee. While this is the short title of the Act, the full title is "ADULTERATION OF COFFEE ACT of 1718 c. 11. An Act against clandestine Running of uncustomed Goods, and for the more effectual preventing of Frauds relating to the Customs" (Mews, Gordon, & Spencer, 1896).

The Act imposed a rather hefty penalty for any

> ...evil-disposed persons who at the time or soon after roasting of coffee, make use of water, grease, butter, or such like material whereby the same is made unwholesome and greatly increased in weight, to the prejudice of His Majesty's Revenue, the health of his subjects, and to the loss of all fair and honest dealers
>
> *(Lely, 1894)*

Here, one can find evidence from over three hundred years ago that the British Parliament had a keen awareness of not only the different times prior to consumption that a food item (in this case, tea) could be adulterated, but also that a range of means exists to do so. Also important is that they not only acknowledge in this act the impact on public health, but also the possible economic impact. This was amended in 1724 with **"AN ACT FOR MORE EFFECTUAL PREVENTING FRAUDS AND ABUSES IN THE PUBLIC REVENUES"** (11 Geo. 1, c. 30). Whereas the earlier Act focused on wholesomeness of the product, it did not address situations where a characteristic other than wholesomeness was at issue. This new amendment addresses not only tea:

> No dealer in tea, or manufacturer or dryer thereof, or pretending so to be, shall counterfeit or adulterate tea, or cause to procure the same to be counterfeited or adulterated, or shall alter, fabricate, or manufacture tea with terra japonica [*also known by many names including "catechu" – it is an extract of acacia trees used variously throughout many centuries as a food additive and dye*] or with any drug or drugs whatsoever, nor shall mix or cause or procure to be mixed with any leaves, other than leaves of tea, or other ingredients whatsoever, on pain of forfeiting and losing the tea so counterfeited, adulterated, altered, fabricated, manufactured, or mixed, and other thing or things whatsoever added thereto or mixed or used therewith, and also the sum of one hundred pounds
>
> *(Lely, 1894)*

But also, coffee was the subject of adulteration that could result in large fines for both roasters, mixers, dealers, and sellers guilty of adulteration if they,

> in order to increase the weight of roasted coffee, ... defraud and impose upon such as bur the same, divers evil-disposed persons, at the time or times of roasting such Coffee, ... do use or mix, or cause to be used or mixed therewith, or do add or cause to be added thereto, butter, lard, grease, water, or other materials, whereby such coffee is rendered less wholesome, to the prejudice of the health of his majesty's subjects, and to the loss and injury of all honest and fair dealers therein: ...
>
> *(Lely, 1894)*

Whereas this new amendment included coffee and increased the penalty amount, it also included much greater depth of clarity in terms of what qualifies as adulterating the commodity. Six years later, this Act was again amended to focus on additional commodities with the ACT TO PREVENT FRAUDS IN THE REVENUE OF EXCISE, WITH RESPECT TO STARCH, COFFEE, TEA, AND CHOCOLATES (1730) (4 Geo. 2, c, 14). This time, the description of the illegal acts included much more clarity:

> whereas several evil-disposed persons do frequently dye, fabricate, or manufacture very great quantities of sloe leaves, liquorish leaves, and the leaves of tea that have been before used, or the leaves of other trees, shrubs, or plants in imitation of tea, and do likewise mix, colour, atain, and dye such leaves, and likewise tea with terra japonica, sugar, molasses, clay, logwood, and with other ingredients, and do sell and vend the same as true and real tea, to the prejudice of the health of his majesty's subjects, the diminution of the revenue, and to the ruin of the fair trader

It was not only this actus reus (bad act) of the crime but also the punishments that illustrate the severity of such adulterations, where:

> the remedy of such frauds and abuses for the future be it enacted ... if any person or persons who shall be a dealer in or seller of tea, ... shall sell or vend or utter, offer, or expose to sale, or shall have in his,

her, or their custody or possession, any such dyed, fabricated or manufactured leaves in imitation of tea, or any such coloured, stained, or dyed leaves, or tea mixed with any of the ingredients before mentioned, or with any other ingredients whatsoever, such person or persons shall respectively, for every pound of such leaves so dyed, fabricated, or manufactured in imitation of tea, and for every pound of such mixed, coloured, stained or dyed leaves or tea forfeit and pay the sum of ten pounds

(Lely, 1894)

Ten Pounds Sterling (10£) from 200 years ago roughly equate thirty U.S. Dollars ($30).[1] Thus, not only was the fine steep, but the crime broadly defined. Here, one can find a wider awareness of the range of ways in which such adulteration is accomplished, a range of when and where such acts are carried out, a range of materials used as adulterants, and a range of means with which to impact victims. This amendment included an emphasis on prohibiting the use of used tea leaves and of imitation ingredients. Also of note is the change in penalty from 100 pounds per offense to 10 pounds per pound of adulterated product found.

In 1773, the British Parliament passed the TEA ACT to save the faltering East India Company from bankruptcy. Essentially, the Act lowered tea tax the East India Company paid to the British government. As a result, the Act also handed the East India Company its very own monopoly on the American tea trade – giving them the competitive advantage of being able to sell tea in the colonies at a lower expense (thus higher profit.) This was a time when tea smuggling into Britain and the colonies was a challenge to the profit – and thus the taxes – on tea.

However, colonists overwhelmingly viewed the Tea Act as an example of tyranny – taxation without representation. Though the Act removed the duty on tea entering England, it left in place an existing duty (tax) on tea entering the colonies. After ships carrying East India Company's tea landed in Boston Harbor, colonists refused to allow them to unload, and demanded that the tea be returned to England. Massachusetts Governor, Thomas Hutchinson, refused to send back the cargo, prompting Patriot leader, Samuel Adams, to organize on December 16, 1773, what is now referred to as the "Boston Tea Party." Over fifty members of the radically anti-British Sons of Liberty impersonated persons from the Mohawk Nation and boarded the British ships, where they dumped overboard an estimated $1.7 million USD (current value) of tea ("Boston Tea Party Damage," 2019).

---

[1] https://www.measuringworth.com/calculators/ppoweruk/

Three years later, Britain again saw more clarification regarding fraud and tea with the **1776 ADULTERATION OF TEA ACT** (17 Geo. 3. c. 29.). This Act provided for "… the more effectual Prevention of the manufacturing of Ash, Elder, Sloe, and other leaves, in imitation of Tea, and to prevent Frauds in the Revenue of Excise in respect to Tea" (Lely, 1894).

While this sounds like a bit of digression into history, this chronology of policy and events offers a glimpse at the importance of tea during that century. Whereas some may call such earlier Acts against fraud and adulteration of tea to be frivolous, as opposed to Acts pertaining to meat or all foods in general, tea held a significant position of importance. Tea became deeply rooted in culture, a symbol of wealth, a trade commodity, and a form of revenue through taxes.

Starting in the mid-1800s, attention on adulteration now focused on food in general. Frederick Filby, in his 1934 book *History of Food Adulteration*, credits the work of German chemist Frederik Accum for having "finally brought the storm over adulteration in 1820…. From that time onward, adulteration has come more and more before public attention" (Filby, 1934). Thus, the push for authenticity of food is an effort with roots that date back to the early 19th century. Accum began his crusade for national adulteration legislation with the publishing of his groundbreaking 1820 writing – "A Treatise on Adulteration of Food and Culinary Poisons." In this, Accum exposed and criticized the food processing industry for "normal" practices – especially the use of chemical additives (Accum, 1820). This work also marked the beginning of public awareness of the need for food safety oversight.

Some thirty years after the publishing of Accum's Treatise, *Lancet*, a British medical journal, published a series of devastating reports in the 1850s on food adulteration, relying on commissioned analyses of food samples (Wilson, 2005). Dr. Arthur Hassall, a British physician and chemist, exposed, among many unsavory practices. For instance, he showed how pickled vegetables were dyed with lead- and copper-based colors. Profit drove adulteration of common items, such as milk (watered), flour (alum added), and beer (lead acetate added) (Clayton, 1908). In their summary of investigations into these incidents, the British Parliament's House of Commons' Committee on Adulteration of Food mostly gave up on the belief that "the forces of competition and the knowledge of the consumer were sufficient to guarantee the sale of unadulterated foodstuffs of adequate quality" (House of Commons, 1872).

Hassall used *Lancet* as a platform to "name and shame" individual shops for the fraudulent products they sold. His work led directly to the passage of the **1860 FOOD ADULTERATION OF FOOD AND DRINK ACT** and later British legislation against these practices (Coley, 2005). The 1860 Act has long been viewed as a failure. Though it involved, for the first time, the appointment of public analysts, very few were appointed, thus the regulation was not generally acted upon.

To reverse this, Parliament passed the ADULTERATION OF FOOD AND DRINK AND DRUGS ACT OF **1872**, which mandated the appointment of public analysts. This new Act designated the sale of mixtures, such as additives along with main ingredients, as illegal unless clearly declared at the point of sale. Also, as evident by the Act's name, the Act covered drugs for first time. This Act described how adulteration caused a "great hurt" to Her Majesty's subjects and endangered their lives (Lely, 1894).

Three years later, THE SALE OF FOOD AND DRUGS ACT OF 1875 made a positive impact on improving food quality, continued the progress of preventing widespread adulteration, and made new efforts to define and regulate food purity. A key element of the Act's effectiveness was its emphasis on strict liability. Another Act that year, THE PUBLIC HEALTH ACT OF 1875, granted powers for food inspection and seizure.

THE SALE OF FOOD AND DRUGS ACT OF **1899** is notable in how it clarified that the definition of food would now include "any article which ordinarily enters into or is used in the composition or preparation of human food." As a result, this Act placed significant focus on food quality and nutrition.

Two years later, in their replacement of the 1872 Adulteration of Food Act with stronger legislation, England's SALE OF FOOD AND DRUGS ACT OF **1875** further defined the term "food," required consumer-driven standards, and set strict liability for food-related offences (Lely, 1894). The **1879** SALE OF FOOD AND DRUGS AMENDMENT ACT introduced changes to solve conflicts in court interpretations and decisions from courts in England and Scotland (Howman, 1901).

The British FOOD AND DRUGS ACT OF **1938** combined food, drugs, and public health legislation relating to food. Specific wording included that the Act "Prohibited the addition of substances to foods so as to render the food injurious to health." For the first time, truth in food labels gained legislation, as the Act authorized policymakers to make regulations governing the labeling of food and set forth penalties for false or misleading labels and advertisements. Further, the UK now designated food poisoning as a reportable communicable disease across country.

Legislation directed to protect consumers from adulteration of specific foods and commodities broadened to a more general approach to all foods and even drugs. These developments in policy reflect changes in behavior – offender, industry, retail, consumer, and legal/political. The many revisions can also be interpreted as a potential weakness – detrimental to food crimes' response.

Of meat and men: A brief history of adulteration policy in the United States

During the U.S. Civil War, President Lincoln signed an Act on May 15, 1862, establishing United States Department of Agriculture (USDA) with a focus mostly on research and discovery. This newly established USDA financed agricultural exploration in foreign lands and hired botanists trained to search for new plants and varieties that would launch new agriculture in the United States. Not until forty years later, however, was the USDA granted authority to regulate and inspect meat. Figure 1.1 summarizes a brief history of food safety within the United States, from the creation of the United States Department of Agriculture in 1865 to the passage of the Food, Drug, and Cosmetic Act of 1938.

More than two decades into the life of the Department of Agriculture, the Commissioner of Agriculture appointed Harvey W. Wiley, MD as chief chemist in 1883 (U.S. Department of Agriculture, 2018). Wiley soon focused his attention and government funding toward the investigation of food adulteration ("Harvey Washington Wiley," 2018). Within four years, Wiley published, at the direction of the Commissioner of Agriculture, a series of Technical Bulletins on Foods and Food Adulterants (U.S. Department of Agriculture, 1887). By the end of the decade, the USDA issued Bulletin 25: "A Popular Treatise on the Extent and Character of Food Adulterations," clearly advocating for national legislation on food adulteration (Wedderburn, 1890).

In a number of unsuccessful attempts between 1897 and 1901, Wiley worked with various organizations to propose various versions of pure-food legislations to Congress ("Harvey Washington Wiley," 2018). Wiley would also go on to experiment with live volunteers, referred to as his "Poison Squad," to determine the effects of preservatives on the human body.

Of Meat and Men Timeline

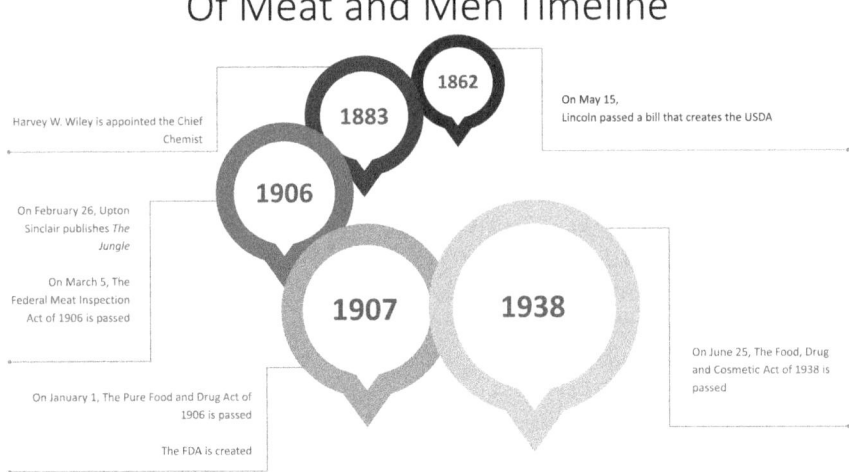

Figure 1.1 This diagram summarizes a brief history of food safety within the United States, from the creation of the United States Department of Agriculture in 1865 to the passage of the Food, Drug, and Cosmetic Act of 1938.

The first significant public call-to-action for "just" food in America came through the landmark reform-oriented work of investigative journalist, Upton Sinclair, who, in 1904, spent seven weeks working undercover in Chicago's meatpacking plants. A year later, he wrote a series of articles for a Socialist political newspaper called *Appeal to Reason*, in which he exposed how unsanitary working conditions were in the plants, and how the meat industry was putting consumers at risk for disease. Sinclair soon published his exposé in the form of his 1906 novel *The Jungle*, allowing the world their first significant notice of the unseen dangers on their dinner plates. Even though Sinclair's intended message was support for socialism, readers paid a great deal of attention to the two chapters in which he described, in detail, the conditions under which meat was prepared.

Readers' concerns soon became a political issue and escalated into a full-blown "meat scandal" in President Theodore Roosevelt's administration. Though initially referred to by Roosevelt as a "Muckraker" for his role as an investigative journalist who exposed a social/corporate ill, Sinclair would later engage directly with the president over the food conditions in Chicago. In one of many letters between Sinclair and President Roosevelt, the author described how the industry in Chicago took steps, after he published his novel, to prevent others from taking a look at what took place inside the processing facilities or, as Sinclair wrote: "The lid is on in Packingtown" (Sinclair, 1906). In response, President Roosevelt sent his own team of commissioners who ultimately proved that the conditions reported by Sinclair were authentic.

Despite the efforts of major meat companies to counter the findings and arguments against the industry, including the Franco-American Food Company's nearly whole page "Open Letter to President Roosevelt and the American Nation" in the *New York Times*, in which they described themselves as the "Packers of Honestly and Cleanly Made" products (The Franco-American Food Company, 1906), the American public was already convinced of the deplorable conditions in the meat-packing industry and were not persuaded by food industry attempts to improve public relations. Consumers across the nation, as well as merchants in many other countries who lost sales due to the "scandal" with bad meat from Chicago, supported strong food safety legislation. This, along with the findings of Roosevelt's investigative commission, no doubt gave strength to the president's decision to sign into law two key piece of food safety legislation.

1. THE FEDERAL MEAT INSPECTION ACT OF 1906 (FMIA) established authority for federal meat inspection.
2. THE PURE FOOD AND DRUG ACT OF 1906 banned the manufacture, sale, or transportation of adulterated or misbranded or poisonous or deleterious foods, drugs, medicines, and liquors and established what would later become the U.S. Food and Drug Administration (FDA).

The Pure Food and Drug Act was later described by R.W. Dunlap, the Acting Secretary of Agriculture, in 1925 as "one of the most beneficent pieces of legislation ever passed by Congress" (Dunlap, 1925).

THE FOOD, DRUG, AND COSMETIC ACT OF **1938**, a subsequent law, repealed some of the 1906 FMIA while empowering the FDA to require food (other than that regulated by the USDA) to conform to three kinds of food standards:

1. Standards (definitions) of identity,
2. Standards of quality, and
3. Standards regulating the fill of container.

No provisions were established by this time, however, for federal inspection by the FDA related to food defense or authenticity, let alone food safety.

From public and government awareness of failures in our food system came new regulations and greater levels of regulatory control. These did not come without criticisms and opposition from some within the food industry. Experts, including Harvey Wiley, the USDA chief chemist 1883–1912, criticized the often-relaxed implementation of new food safety regulations.

In 1925, Wiley wrote a letter to President Calvin Coolidge in which he addressed "a 'shocking' neglect on the part of the United States Government to enforce the Food and Drugs Act, for which I labored incessantly (sp) for twenty-five years" (Wiley, 1925). He did not technically send the letter to the President, as he published the letter in *Good Housekeeping Magazine*, then the director of the Bureau of Foods, Sanitation, and Health for the magazine. Wiley criticized the government for having often turned a blind eye on specific cases that appeared to violate the law, and he discussed how "failures to administer the law" by those superior to him were so "shocking" that he ultimately retired voluntarily. Wiley noted in his letter that "The proper enforcement of the Food and Drugs Act is intimately related to the public health," but offered his assessment in that "the health and efficiency of our citizens are continually threatened" (Wiley, 1925). Though Wiley's criticisms may have been justified, food safety had come a long way since the turn of the century.

Taking a summary review, England's **1860 ADULTERATION ACT** and America's **1906 PURE FOOD AND DRUG ACT** are two of the earliest pieces of legislation to provide generalized regulation of food and drugs on a national scale. In both the European and American events, political landscapes conducive to reform in protecting consumers and becoming modern regulatory states came about through the hard work of individuals who campaigned for legislation to prevent adulteration. Legislative changes came about after the hard work of investigative journalists who were enthusiastic to bring the evils of adulteration to the forefront of the public mind, and the ignited demands from consumers.

Over the next several decades, new sciences, new food production technologies, and new consumer trends, demands, and behaviors would collide to erode the level of food safety and reverse some of the progress that had been made. Frozen foods and improved means of transportation allowed for raw ingredients and prepared foods to last longer and travel farther. At the same time, mid-century Americans were gradually beginning consuming more meals outside the home and, along with Americans' passion for driving, fast food restaurants eventually grew in numbers and popularity, even securing their place as part of the "American culture."

Modern progress on food fraud and food defense: Food poisoning and bioterrorism

In the mid-1980s, some real threats emerged as followers of Bhagwan Shree Rajneesh, a spiritual teacher in India, carried out "the first and largest bioterrorism attack in the United States, of Food Poisoning/Bioterrorism on American soil" (Powell, 2018). The "1984 Rajneeshee Incident" resulted in 751 recorded illnesses, 45 hospitalizations, and no deaths of citizens in The Dalles, Oregon, from *Salmonella enterica* Typhimurium (Detwiler, 2016).

A lengthy investigation by the Oregon Health Department and various other local and federal authorities found that the perpetrators spread liquid tainted with the *Salmonella* pathogen on surfaces in many public places, including the Wasco County Courthouse. Perpetrators introduced the pathogen into the drinking water, salad bars, and even salad dressings at nearly a dozen restaurants in The Dalles, Oregon. Their purpose was determined to be as insidious as the act itself – to incapacitate voters in order to influence the outcome of a local 1984 election in their favor, placing cult members, known as Rajneeshees, into office. This unprecedented act of terrorism would eventually force policy makers to focus on food defense in the form of "defining the illegality of ill-intended use, production, dissemination, or storage of biological agents" (Ryan & Glarum, 2008).

Food defense became a priority after the 1984 Rajneeshee bioterror attack and after the attacks on September 11, 2001. Not even a year had passed since the 9/11 attacks on New York's World Trade Center buildings and on the Pentagon, outside of Washington, DC, when the World Health Assembly, the decision-making body of the World Health Organization (WHO), adopted a 2002 resolution expressing serious concern about threats against civilian populations by deliberate use of agents disseminated via food.

Within a year, WHO published "Terrorist Threats to Food: Guidance for Establishing and Strengthening Prevention and Response Systems" – a food safety/food terrorism document for national government policy makers (World Health Organization, 2003). Focusing on food, food ingredients, and water (in the forms of food ingredients and of bottled water), the document classifies food safety as an essential element of modern, global public health security. It goes on to define "food terrorism" as:

> an act or threat of deliberate contamination of food for human consumption with biological, chemical, and physical agents or radionuclear materials for the purpose of causing injury or death to civilian populations and/or disrupting social, economic or political stability.

In outlining the potential effects of food terrorism, the WHO utilized data from "unintended" foodborne disease outbreaks to describe the toll of potential disease and death. The document looked at how a single incident of "unintentional contamination" of just one kind of food could infect hundreds of thousands of people with a "serious debilitating disease." Then, the WHO goes on to extrapolate the effects of some more deliberate and dangerous attack on our food supply.

The impact on trade and the economy is discussed as a "primary motive" for food terrorism. Recalls in American markets of foreign fruits resulted in bankruptcy of international growers and shippers after consumers around the globe shunned such products. The WHO document details specific events in recent history when individual U.S. recalls of domestic ground beef contaminated with *E. coli* O157:H7 and lunch meats contaminated with *Listeria* numbered in the 20 millions of pounds of affected product each.

The USDA's Food Safety Inspection Service (FSIS) lists on its webpage a great amount of information online for each recall issued in the United States. The number of entries for individual recalls is staggering. Not only are the examples listed by the WHO the tip of the iceberg in terms of the numbers of recalls and the quantity of food products adulterated, but a look at data from the Bureau of Labor Statistics shines more light on scope of this economic impact. When analyzing Consumer Price Index average price data specific for the products and the year of the recalls, one learns that the approximate dollar value loss of just the two beef recalls listed in the WHO document come in at $44 million and $61 million, respectively.

Again, the WHO points to the significant financial impact on the market and related stakeholders. Beyond the loss of profit and the closing of businesses and the financial toll on individual countries, however, the WHO uses lessons learned from outbreaks and recalls over the past twenty years to emphasize that foodborne diseases have the potential of causing the disruption of global trade and economic stability and may even impact political stability.

While the WHO published "Terrorist Threats to Food" to provide member governments with guidance on preventing the deliberate contamination of food, some of this document's main points hold significant meaning for unintentional food problems. The understanding of those in the industry of every facet of the food chain, from farm to table, is critical in identifying and preventing failures and violations of the system.

The United States House of Representatives introduced the Public Health Security and Bioterrorism Preparedness Response bill as H.R. 3448 on December 11, 2001. It passed the House the next day almost unanimously, then passed the Senate unanimously on December 20, 2001, as THE PUBLIC HEALTH SECURITY AND BIOTERRORISM PREPAREDNESS RESPONSE ACT OF **2002** – an Act "To improve the ability of the United States to prevent, prepare for, and respond to bioterrorism and other public health emergencies." Surrounded by leaders from the Department of Health and Human Services, and the USDA President George W. Bush signed the Act into law as Public Law 107–188 on June 12, 2002.

The Act established procedures for preparation for bioterrorism and public health emergencies, as well as the National Disaster Medical System, comprised of teams of health professionals. Further, the rules under this Act include security risk assessment of individuals who have access to the select agents and toxins, with the purpose being to restrict access from any person who meets the criteria of a "restricted person" as defined in the **USA** PATRIOT ACT OF **2001** (PL 107-56 – October 26, 2001) signed into law by President George W. Bush the previous year.

A subpart of the Public Health Security and Bioterrorism Preparedness Response of 2002, the **Agricultural Bioterrorism Protection Act of 2002** (80 FR 10627, 7 CFR 331, 9 CFR 121), provides for the regulation of certain biological agents and toxins that have the potential to pose a severe threat to human, animal, and plant health, or to animal and plant products.

Under this Act, 67 select agents and toxins as per 7 CFR Part 331, 9 CFR Part 121, and 42 CFR Part 73 are categorized into three areas. First, those subject to regulation by Health and Human Services (HHS) – specifically the Centers for Disease Control and Prevention (CDC), such as SARS-associated coronavirus (SARS-CoV), Ricin, Ebola, and South American Hemorrhagic Fever viruses. Second are those subject to regulation by the USDA – specifically the Animal and Plant Health Inspection Service (APHIS), such as Avian influenza virus, swine fever virus, and USDA plant protection and quarantine agents and toxins. Finally – those that overlap and are subject to regulation by both agencies, such as various *Bacillus* and *Brucella* agents and Venezuelan equine encephalitis virus ("Select Agents and Toxins List," 2017).

Over the past two decades, food authenticity has grown slowly as a concern in the United States, when fraud within certain commodities (e.g., honey and seafood) began capturing headlines. Major international events, such as the 2008 melamine scandal in China (when adulterated milk and infant formula caused over 300,000 illnesses, some 54,000 infants hospitalizations, and the deaths of six babies (Macartney, 2008), as well as the 2013 horsemeat scandal in Europe prompted new laws and new agencies to investigate and prevent. The huge scandal rocked Europe in 2013, however – the horsemeat incident where horse DNA was found in process beef products – triggered a change far different than simply new laws.

Within a year of the 2013 horsemeat scandal, the UK decided that something needed to be done and that the alignment of laws, regulations, and real industry oversight needed to be reviewed. So they commissioned an external expert, professor Chris Elliott, Chair of Food Safety and Microbiology at Queen's University Belfast, to conduct a government review of kind of what went wrong, to look at ways in which we can change the way we operate to prevent it happening and also to deal with when these things do happen in future in, perhaps, a more efficient manner. One recommendation that Professor Elliot made was that the UK would have its own dedicated unit to deal, not with regulatory noncompliance, which is the bulk of offending within the food world, but with that very narrow scene of fraud – serious dishonesty within food supply chains. The UK's National Food Crime Unit, headed by Andy Morling, grew rapidly from a few employees to eighty members of staff and modern law enforcement agents. The unit works not only to deter offenders from committing crimes through food, but also has dedicated teams whose sole focus is working with industry to help them design out vulnerabilities within their supply chains and to look at ingenious new ways to reduce their vulnerability.

Defining modern food reputations

Today, food industry leaders and policymakers focus on multiple separate, yet overlapping efforts to protect and sustain our food supply. These approaches to keeping our food safe and abundant have been rather piecemeal. Food quality, food safety, food fraud, food defense, and food security have become the modern food reputation approaches that consumers now take into consideration (and policymakers are now faced to contemplate).

Food quality

A variety of definitions for food quality exist, with next- or end-use consumer typically having final definition. Common to most definitions, is that food quality involves processes to support consistent manufacturing specifications and consumer's desired, premium end product (density, color, smell, texture, viscosity, etc.). The perception of food quality defined by the consumer adds to the concerns related to food waste by assuming that only fresh and quality fruits and vegetables are perfectly unblemished. For a scientist, food quality could be determined by genetic make-up and nutritional content (Schulz & Kopke, 1997).

In short, food quality can be described as characteristics that make it acceptable to consumers' perceived values or priorities in the following properties:

- **Appearance**: such as color or even packaging characteristics
- **Smell**: such as perceived "freshness"
- **Taste**: such as salty or savory
- **Touch**: such as texture

Measures to determine the authenticity of food quality date back to the earliest days of bartering and markets and organoleptic testing (use of sensory organs for evaluations of the odor, flavor, and texture of food). These measures served for nearly a century as the foundation of the USDA's work in detecting foodborne pathogens before the advancement of modern methods (such as whole genome sequencing.) Food quality determination is nothing new, only the means of affecting quality and the science of detection have changed.

Food safety

In their 2015 report on the Global Burden of Foodborne Diseases, WHO states that each year as many as 600 million, or almost one in ten people in the world, fall ill after consuming contaminated food. Of these, 420,000 people die, including 125,000 children under the age of five (World Health Organization, 2015). The U.S. CDC estimates that each year as many as 48 million Americans become ill from foodborne pathogens, 128,000 are hospitalized, and 3,000 die (Centers for Disease Control and Prevention, 2014). An important observation is that a large portion of foodborne illness cases could be prevented (Mead, et al., 1999).

In their fact sheet on food safety, WHO discusses it in terms of preventing foodborne diseases, defining foodborne illnesses as being:

> … usually infectious or toxic in nature and caused by bacteria, viruses, parasites or chemical substances entering the body through contaminated food or water. Foodborne pathogens can cause severe diarrhea or debilitating infections including

> meningitis. Chemical contamination can lead to acute poisoning or long-term diseases, such as cancer. Foodborne diseases may lead to long-lasting disability and death. Examples of unsafe food include uncooked foods of animal origin, fruits and vegetables contaminated with feces, and raw shellfish containing marine biotoxins
>
> *(WHO, 2019)*

Looking to the Foundation Food Safety System Certification (FSSC) 22000 scheme, more elements that define food safety can be found.

> **Food/Feed Safety** – "The policies, processes and procedures, materials, facilities, and monitoring systems applied to Food or Feed products to ensure they will not cause harm to humans or animals or adversely affect their health when utilized according to their intended purpose" (FSSC-0-006.2).
>
> **Food/Feed Safety Hazard** – "Biological, chemical, physical agent or allergen in food/feed, or condition of food/feed, with the potential to cause an adverse health effect to humans and/or animals" (FSSC-0-006.2/ISO 22000, section 3.3).

From this, food safety risks and hazards can be defined simply as relating to or stemming from:

- **Biological contamination:** such as *E. coli, Salmonella, Listeria,* etc.
- **Physical contamination:** such as glass, rocks, plastic, etc.
- **Chemical contamination:** such as pesticide residuals, allergens, and other pollutants or toxic substances

The **FDA Food Safety Modernization Act of 2011** (Pub. L. 111-353), commonly referred to as "FSMA," aims to "reduce risk of illness attributed to food from facilities subject to preventive controls rule under the act" (Milazzo, 2015). This public law is a modification to the **1938 Federal Food, Drug, and Cosmetic Act** (21 U.S.C. 301 et seq.). Specific sections of FSMA focus on improving "capacity to prevent food safety problems," "capacity to detect and respond to food safety problems," and "the safety of imported foods." In passing this Act, Congress directed the FDA to coordinate inspection and compliance efforts through state agencies and resources.

The investigation and reporting of foodborne illnesses by state and county health departments are critical in the prevention of foodborne disease in the United States (Lynch, Painter, Woodruff, & Braden, 2006). However, the detection of hazards is second to the mitigation of food safety failures in preventing adverse impacts on the health of humans or animals. Consistent and widespread determination of safe food throughout the production process would serve as an effective deterrent to food fraud and to the risks for which food defense measures are designed. The application of food safety verification activities plays a significant role in determining if acts and crimes involving food should be defined as food fraud, a breach in food defense, or of risk in food security, and would, thus, continue to serve the entire purpose of preventing harm to consumers.

Technically, all among its foundational rules, FSMA includes the following:

- Current Good Manufacturing Practice and Hazard Analysis and Risk-Based Preventive Controls for Human Food
- Current Good Manufacturing Practice and Hazard Analysis and Risk-Based Preventive Controls for Food for Animals
- Sanitary Transportation of Human and Animal Food
- Standards for the Growing, Harvesting, Packing, and Holding of Produce for Human Consumption (also known as the "Produce Safety Rule")

Other rules do play a role in food safety, but, perhaps, not necessarily as obvious:

- Accredited Third-Party Certification
- Foreign Supplier Verification Programs (FSVP)
- Mitigation Strategies to Protect Food Against Intentional Adulteration
- Voluntary Qualified Importer Program (VQIP)

Foodborne illness affects people of all ages, but certain "vulnerable populations," such as very young children, pregnant women, the elderly, and people with a compromised immune system are more likely to develop severe symptoms or even die. When families are sitting at the bedsides of those who become ill or next to the deathbed of someone who dies from food that is not safe, they do not care about a legal definition of the act or failure that may or may not come until many months later.

Food fraud

Looking to the Foundation Food Safety System Certification (FSSC) 22000 scheme, Food Fraud Prevention is defined as "The process to prevent food and feed supply chains from all forms of economically motivated, intentional adulteration that might impact consumer health" (FSSC-0-005.1).

The FDA FSMA final rule for Mitigation Strategies to Protect Food Against Intentional Adulteration defines Mitigation Strategies to Protect Food Against Intentional Adulteration as those "aimed at preventing intentional adulteration from acts intended to cause wide-scale harm to public health, including acts of terrorism targeting the food supply. Such acts, while not likely to occur, could cause illness, death, economic disruption of the food supply absent mitigation strategies."

The National Center for Food Protection and Defense, in 2011, defined food fraud in terms of seven different risks:

- **Adulteration:** mixing matter of an inferior and sometimes harmful quality with food intended to be sold. As a result, it becomes impure and unfit for human consumption (A.K.A. Dilution).
- **Tampering:** legitimate product and packaging are used in a fraudulent way (A.K.A. Misbranding).
- **Overrun:** legitimate product is made in excess of production agreements.
- **Theft:** legitimate product is stolen and passed off as legitimately procured.
- **Diversion:** the sale or distribution of legitimate products outside of intended markets.
- **Simulation:** one product is designed to look like real, labeled product (A.K.A. Substitution).
- **Counterfeiting:** intellectual property rights infringement, including fraudulent product/packaging.

Almost every food fraud risk listed demands vigilant actions regarding food safety. Impact on public health typically takes place long before any county health official, federal regulator, or court will determine whether an act is intentional, economically motivated, or with true *mens rea*. Though the FDA's FSMA rules include one on Mitigation Strategies to Protect Food Against Intentional Adulteration, FDA's former Deputy Commissioner for Foods and Veterinary Medicine, Stephen Ostroff, MD, stated at a 2016 food industry conference that the FDA will only focus on food fraud when it affects food safety (Ostroff, 2016). Unfortunately, before any fraudulent or other intentional act gains classification is when such acts are identified – typically after public health threats have become a reality. As a result, regulators – especially as the mission of the FDA Food Safety Modernization Act (FSMA) spells for – must focus on authenticity to "reduce risk of illness attributed to food from facilities subject to preventive controls rule under the act" (Milazzo, 2015).

Specific sections of FSMA focus on improving "capacity to detect and respond to food safety problems," and "the safety of imported foods," but, perhaps more importantly, the "capacity to prevent food safety problems." Authenticity is a key component of FSMA's mission beyond the intentional adulteration rule, including both preventive controls rules and even the foreign supplier verification rules.

Food defense

Looking to the Foundation Food Safety System Certification (FSSC) 22000 scheme, food defense is defined as processes to prevent "… food and feed supply chains from all forms of ideologically or behaviorally motivated, intentional adulteration that might impact consumer health" (FSSC 22000-0-005.2). Additional definition for industry can also be found as "preventive measures … The organization shall put in place appropriate preventive measures to protect consumer health impacts" (FSSC-2-006.1: 2.1.4.5 Food Defense 2.1.4.5.2).

The National Center for Food Protection and Defense, in their 2011 backgrounder on "Defining the Public Health Threat of Food Fraud" defined food defense in terms of three different risks:

- **Industrial sabotage:** intentional contamination by an insider or competitor to damage the company, causing financial problems/recall but not necessarily to cause public harm.
- **Terrorism:** the reach and complexity of the food system has caused concern for its potential as a terrorist target.
- **Economically motivated adulteration:** acts against a product for the purpose of increasing the apparent value of the product or reducing the cost of its production (i.e., for economic gain).

Similar to food fraud, the point at which any intentional act gains classification as one in the category of food defense is typically after public health threats have become a reality. Regular, sustained actions to mitigate food defense crime serves as a deterrent as much as it serves to protect public health.

The December 2016 fake infant formula incident in New Zealand serves as, perhaps, the most recent example of authenticity measures identifying an economically motivated activity – against food intended for babies – taking place on an international scale (ties to a Russian counterfeiting ring) (Theunissen, 2016). The global concerns regarding food defense and infant formula are not without merit. The 2008 China melamine in baby formula scandal resulted in 300,000 victims, 54,000 babies hospitalized, and 12 deaths (Huang, 2014). Further, the impact of the 2008 incident on the Chinese economy, due to the loss of confidence from consumers in China and abroad, cannot be ignored.

Food security

The United Nations (UN) Food and Agriculture Organization (FAO) 2009 document titled *Declaration of the World Food Summit on Food Security* discusses food security in that it "…exists when all people, at all times, have physical, social and economic access to sufficient, safe and nutritious food which meets their dietary needs and food preferences for an active and healthy life." The declaration also defines food security in terms of four different risks:

- **Availability**: the supply of food through production, distribution, and exchange.
- **Access**: the affordability and allocation of food, as well as the preferences of individuals and households.
- **Utilization**: the metabolism of food by individuals – can be affected by food safety, nutritional values, food choice, and cultural preferences.
- **Stability**: the ability to obtain food over time, as food insecurity can be transitory, seasonal, or chronic. This can be impacted by failures in food sustainability, defense, etc.

The "utilization" component of food security depends on authentication for food safety, nutritional values, food choice, and cultural preferences. Foods pertaining to modern diets, such as vegan, vegetarian, allergy-conscious, impacted immune status, etc., require food authenticity from retail and their suppliers. People experiencing poverty, weather phenomenon such as droughts or floods, conflict, forced migration, or terrorism could be at risk for insufficient access to proper foods necessary to feed themselves and their families.

The FAO uses a definition of food security that comes from the 1996 World Food Summit. "Food security exists when all people, at all times, have physical and economic access to sufficient, safe and nutritious food that meets their dietary needs and food preferences for an active and healthy life" (FAO, 2006).

A look at food security should not go without considering how the UN would interpret the larger context of food reputation approaches. In 2015, the UN created "Transforming our World: The 2030 Agenda for Sustainable Development," which focuses on seventeen sustainable development goals (SDGs). In addition to improving lives, caring for the planet, and stopping climate change, these goals include building partnerships between all member states to rid the world of poverty by improving education, health, and equality.

The SDGs are interdependent and interrelated, thus, reaching targets for one goal inevitably improves other goals. The five food reputations have relationships with the SDGs. They are easily aligned with food safety and food security (such as "Good Health and Well-being," "Clean Water and Sanitation," and "Sustainable Cities and Communities.") Similarly, many alignments can be found between the SDGs and both food fraud and food defense – specifically "Zero Hunger," "Good Health and Well-being," "Clean Water and Sanitation," "Industry, Innovation, and Infrastructure," "Responsible Consumption and Production," and "Peace, Justice, and Strong Institutions" (United Nations, 2015).

Crimes related to food fraud and food defense

According to Andy Morling, "Food crime is quite simply a crime that makes victims out of all of us, the young, the old, the rich, the poor, the discerning and the less discerning" (Morling, 2017). Morling, who joined the UK's Food Standards Agency in 2015 as Head of Food Crime with the task of building and leading the Agency's National Food Crime Unit "after the 2013 horsemeat scandal to protect the public from this newly synthesized form of serious crime" (Morling, 2017).

Morling views food crime as a crime problem rather than a food problem. Speaking before a 2017 National Center for Food Protection and Defense Conference in Minneapolis, MN, Morling argued that not only is there not any one recognized or agreed upon definition of a food crime, but that one is not needed "because many hours, many days had been spent in pursuit of the perfect definition….and, this has been "to the detriment of the response [to food crimes]" (Morling, 2017). This is evident by how many amendments to the early adulteration Acts can be found as far back as the 18th century in Britain.

In a broader sense, Morling describes food crime as "often sophisticated" and always involving dishonesty – "one of the key things that distinguishes food crime from regulatory noncompliance." It typically involves a seller "misrepresenting one facet or another of a food product to a buyer …its composition, its durability, its country of origin, its production method, or its health properties…."

Morling's deeper definition of a food crime is important to consider:

> It can be subtle like the adulteration of a single food ingredient, or it can be the gross substitution of an entire product with something of lesser value or that's imported. Offenders might seek to profit from a mass market product sold to thousands and thousands of consumers or they may seek to profit from a high value item, a niche item sold direct to a few. Food crime is generally business to business. But in almost every case, the detriment from that fraud rests with the loss suffered by consumers.

> When a business is the victim of fraud from one
> of its suppliers, that retailer invariably – and in all
> instances – passes that detriment on to consumers
> *(Morling, 2017)*

U.S. food company federal fines

The landmark 1993 *E. coli* outbreak tied to undercooked, contaminated hamburger meat sold at Jack in the Box restaurants, killed 4 children and infected 732 people in several states. The Washington State Department of Health found the restaurants to be in violation of WA State minimum cooking temperature. Neither the state nor federal governments filed any criminal charges against the company or any executives, though the company admitted guilt in violating WA State law. The company and several other parties paid multiple out-of-court settlements (Detwiler, 2020).

In 1996, an *E. coli* outbreak, tied to Apple Juice sold by Odwalla, Inc., killed a 16-month-old girl and sickened seventy in several states and Canada (Centers for Disease Control and Prevention, 1996). The U.S. Department of Justice (DOJ) filed sixteen federal criminal counts against the company for distributing adulterated juice. Odwalla ultimately pleaded guilty and agreed to pay a $1.5 million fine and serve a court-supervised probation for five years. No individuals received jail time for their role in this outbreak (Department of Justice, 2009). This was, however, the first criminal conviction in a large-scale food-poisoning outbreak and, at the time, the $1.5 million penalty was the largest criminal penalty in a food poisoning case.

The 2006–2007 *Salmonella* multistate outbreak, tied to contaminated peanut butter, sickened over 700 people in 44 states (Centers for Disease Control and Prevention, 2007). Eight years later, ConAgra Foods agreed to plead guilty to federal charges of Introducing an Adulterated Food into Interstate Commerce, 21 U.S.C. §§ 331(a) and 333(a)(1) and pay $11.2 million in fines. ConAgra also signed a plea agreement admitting that it unknowingly introduced Peter Pan and private label peanut butter contaminated with *Salmonella* into interstate commerce (Department of Justice, 2015). This is the first reported outbreak of a foodborne illness caused by peanut butter consumption in the United States. This was also the largest fine at the time in a food safety case.

In January 2020, Chipotle Mexican Grill Inc., was fined $1.37 million not for violating food laws, but for violating Child Labor Laws (Coley, 2020). The Attorney General for the Commonwealth of Massachusetts announced a series of violations by the company's restaurants in Massachusetts, including:

- Minors working without valid work permits.
- Minors working too late in the evening.

- Minors working too many hours per day and per week (more than the nine-hour daily limit and 48-hour weekly limit).
- Failure to properly notify employees of their rights relating to sick time.
- Failure to provide the attorney general's office with timekeeping records upon request.
- Failure to pay workers within six days of the end of a pay period [based on audits conducted between 2015 and 2019].

This nonfood related fine is notable due to the ongoing investigation in California over the company's failure to notify the local health department about sick employees involved in the company's 2015 Norovirus outbreak at a restaurant in Simi Valley, about an hour north of Los Angeles in Ventura County, California. Five years later, in April 2020, Chipotle Mexican Grill Inc. agreed to pay $25 million in fines to resolve charges related to that one in 2015 and several other single- and multistate outbreaks of various pathogens between 2015 and 2018 (Department of Justice, 2020a; Marler, 2020). Collectively, these outbreaks sickened more than 1,100 people. The federal charges related to violations of the Federal Food, Drug, and Cosmetic Act (21 U.S.C. Ch. 9 § 301 et seq.). Chipotle also agreed to a three-year deferred prosecution agreement that will allow it to avoid conviction if it complies with an improved food safety program (Marler, 2020). The $25 million criminal fine is the largest fine ever in a food safety case.

Barely a week after the DOJ announced the agreement with Chipotle Mexican Grill Inc., they announced another agreement and forthcoming legal actions with Blue Bell Creameries (Department of Justice, 2020b). This agreement stems from the 2015 Listeriosis outbreak, tied to ice cream, frozen yogurt, sherbet, and frozen snacks sold by Blue Bell Creameries, based in Texas, which sickened at least ten in four states, killing three people. The company agreed to plead guilty to two misdemeanor counts of distributing adulterated ice cream products as well as to Civil False Claims Act allegations regarding ice cream products manufactured under insanitary conditions and sold to federal facilities. They will pay a criminal fine and forfeiture amount totaling $19.35 million – an amount that constitutes the second largest-ever amount paid in resolution of a food safety matter. Further, the DOJ charged Blue Bell's former CEO and president, Paul Kruse, with seven felony counts (conspiracy and six counts of wire fraud and attempted wire fraud) related to his alleged efforts to conceal from customers what the company knew about the *Listeria* contamination – a possible twenty-year federal prison sentence and a $250,000 fine for each charge. In July 2020, all charges against Kruse were dropped due to the federal District Court lacking subject matter jurisdiction and DOJ not obtaining a Grand Jury indictment or waiver from Kruse.

U.S. food company trials in federal court

Food fraud and food defense are not alone in their association with crime. Crimes can and have been committed that involved other food reputations. However, beyond convictions for Introducing an Adulterated Food into Interstate Commerce [21 U.S.C. §§ 331(a) and 333(a)(1)], the more notable, even landmark trials of corporate executives and owners of food companies are mostly related to larger crimes that are not specific to food, including Aiding and Abetting (18 U.S.C. § 2), Obstruction of Justice (18 U.S.C. § 1503), Fraud (in general – not specifically "food fraud"), and Conspiracy (18 U.S.C. § 371).

Three examples:

1. *U.S. v. Eric Jensen and Ryan Jensen* (2013) involved the 2011 *Listeria* outbreak tied to improperly cleaned cantaloupe from their Colorado farm. Ranking among the deadliest U.S. outbreaks, this outbreak sickened 147 people in 28 states, resulting in 33 deaths (Centers for Disease Control and Prevention, 2012). The Jensen brothers pled guilty to six counts, including Introducing an Adulterated Food into Interstate Commerce [21 U.S.C. §§ 331(a) and 333(a)(1)] and Aiding and Abetting (18 U.S.C. § 2). In 2014, they received five years of probation, six months of home detention, 100 hours of community service, and were ordered to pay a total restitution of $300,000 to victims' families (U.S. Department of Justice, 2013B).

2. *U.S. v. Quality Egg, LLC* (2014) involved the 2010 outbreak of *Salmonella* that sickened nearly 2,000 consumers nationwide and resulted in the recall of over one half of one billion eggs. The defendants, Austin DeCoster and his son Peter DeCoster, and the company pled guilty in June 2014 to Introducing an Adulterated Food into Interstate Commerce [21 U.S.C. §§ 331(a) and 333(a)(1)]. In 2015, the court sentenced the two owners to serve six-month jail terms, pay a $100,000 fine each, as well as pay restitution to victims. The court placed the corporation on three years' probation and ordered it to pay a fine of $6.79 million (U.S. Department of Justice, 2015A). The two convicted felons appealed all the way to the U.S. Supreme Court, however, the U.S. Supreme Court denied the DeCosters' petition for appeal (Flynn, 2018).

3. *U.S. v. Stewart Parnell, Michael Parnell, Samuel Lightsey, and Mary Wilkerson* (2014) involved the 2008–2009 *Salmonella typhimurium* multistate outbreak resulted in 714 illnesses reported across 46 states, nine deaths tied to contaminated peanuts from the Peanut Corporation of America (PCA) (Centers for Disease Control and Prevention, 2009). The outbreak involved a lengthy investigation by the FDA and the DOJ, as well as a number of

recalls involving more than 3,900 different kinds of products (such as peanut crackers, ice creams, candies, cake mixes, and per foods) from over 360 companies. The USDA described this recall as one of the largest food recalls in U.S. history (Wittenberger & Dohlman, 2010).

While the U.S. DOJ charged Stewart Parnell (the owner), Michael Parnell (PCA's broker), Mary Wilkerson (PCA's QA manager), as well as Samuel Lightsey and Daniel Kilgore (two PCA plant managers) with the violation of Introducing an Adulterated Food into Interstate Commerce [21 U.S.C. §§ 331(a) and 333(a)(1)], the majority of the federal charges included those under Crimes and Criminal Procedure (18 U.S.C. Title 18) – specifically Obstruction of Proceedings before departments, agencies, and committees (18 U.S.C. § 1505), Attempt and Conspiracy (18 U.S.C. § 1349), Fraud by wire, radio, or television (18 U.S.C. § 1343), and Frauds and Swindles (18 U.S.C. § 1341). None of the charges beyond "Introducing an Adulterated Food into Interstate Commerce" specifically or uniquely pertained to food (U.S. Department of Justice, 2013A).

A federal jury convicted Stewart Parnell and his brother Michael of multiple counts of conspiracy, mail and wire fraud, and the sale of misbranded food. Stewart Parnell was also convicted of Introducing an Adulterated Food into Interstate Commerce [21 U.S.C. §§ 331(a) and 333(a)(1)], while he and Wilkerson (the QA manager) were also convicted of Obstruction of Proceedings before departments, agencies, and committees (18 U.S.C. § 1505). At the time of the September 21, 2015 sentencing, U.S. District Judge, W. Louis Sands, stated before the defendants, their families, and the families of those sickened and killed that:

> We place faith that no one would intentionally ship products to market that are contaminated. Striking and strong testimony was heard today. Consumers are at the mercy of food producers for the safety of the products. These acts [of the convicted PCA executives] were driven by profit and the protection of profit - thus greed
>
> *(Detwiler, 2015)*

The owner, Stewart Parnell, was found guilty of all but one of the sixty-eight felony counts with which he was charged, then, later, sentenced to serve twenty-eight years in federal prison. His brother, the broker, was sentenced to serve twenty years in federal prison. The QA manager was sentenced to serve five years in federal prison (U.S. Department of Justice, 2015B).

Food fraud and defense vs. "intentional adulteration"

THE FDA FOOD SAFETY MODERNIZATION ACT (FSMA) final rule for Mitigation Strategies to Protect Food Against Intentional Adulteration is aimed at "preventing intentional adulteration from acts intended to cause wide-scale harm to public health, including acts of terrorism targeting the food supply." The guidance describes such acts as "not likely to occur," while noting that they could still "cause illness, death, economic disruption of the food supply absent mitigation strategies." Of note is that this FSMA rule does not target any specific foods or hazards but "requires mitigation (risk-reducing) strategies for processes in certain registered food facilities" (U.S. Food and Drug Administration, 2018).

As a result, this FSMA rule on intentional adulteration may apply to food fraud and food defense issues – but only if they cause wide-scale harm to public health. Again, the FDA's former Deputy Commissioner for Foods and Veterinary Medicine, Stephen Ostroff, MD, stated at a 2016 food industry conference that the FDA will only focus on Food Fraud when it affects food safety (Ostroff, 2016). One can interpret that this position of the FDA would apply to food crime in the larger context as well.

Food fraud and defense cannot be viewed as unrelated to each other nor to food quality, safety, and security (the rest of the five reputations). If a distributor, retailer, or restaurant has doubt or finds indicators that any of the food reputation approaches cannot be authenticated, then deficiencies involving transparency and traceability are often blamed. If authenticity in many forms is determined and strengthened through transparency and traceability early on in the supply chain, adverse food events can become more preventable. This same form of mitigation applies to all of the five food reputation approaches.

The ability to authenticate food reputation approaches starts with defining food. The approach to defining food must be consistent and emanate from one single recognized global source. With a definitive and agreed upon definition as to the composition of each different kind of food ingredient in the world, trade can be enhanced, new food sources can be developed, and demand can be satisfied. While some have viewed existing definitions of these five approaches in terms of intent, such as in "Food Protection Risk Matrix" (Spink & Moyer, 2011) – another, perhaps more public health-focused view, could be through the focus of authenticity. Food authenticity aims to provide assurance that foods are as they are advertised. Food authenticity does not focus on the ethical, economic gains, or intents behind the act. Further, the construct of an authenticity approach is attached to several, much earlier points in time from farm-to-fork.

Once the definitions for each different kind of food are published, existing technologies can be applied, and new technologies can be developed to detect failures in any and all of the reputations. As a result, significant technological advances, such as blockchain (an immutable database system), must be seen as being strong in one area only if it is strong in supporting all of the food reputations. In a sense, a blockchain for the food reputations is only as strong as its weakest link.

Though the American food supply has become global and consumers benefit from advances in food manufacturing, food distribution, and food policies, the approaches to keeping food safe and abundant have been rather piecemeal – developed at different times. Focusing on food fraud and food defense alone will not be enough to protect consumers. All five different food reputation approaches must be considered as one harmonized and standardized system for the entire world.

References

"Boston Tea Party Damage." (2019). The Boston Tea Party Ships & Museum. Retrieved from https://www.bostonteapartyship.com/boston-tea-party-damage

"Food Safety." (2019, June 4). World Health Organization Fact Sheet. Retrieved from https://www.who.int/news-room/fact-sheets/detail/food-safety

"Food Security." (2006, June). Food and Agriculture Organization Policy Brief Issue 2. Retrieved from http://www.fao.org/fileadmin/templates/faoitaly/documents/pdf/pdf_Food_Security_Cocept_Note.pdf

"Harvey Washington Wiley." (2018, January 10). Science History Institute. Retrieved from https://www.sciencehistory.org/historical-profile/harvey-washington-wiley

"Select Agents and Toxins List." (2017). Federal Select Agent Program. CDC, USDA. Retrieved from https://www.selectagents.gov/SelectAgentsandToxinsList.html

Accum, F. (1820). A Treatise on Adulteration of Food and Culinary Poisons. London: Longman, Hurst, Rees, Orme, and Brown. Republished electronically on The Public Domain Review. Retrieved from https://publicdomainreview.org/collections/a-treatise-on-adulteration-of-food-and-culinary-poisons-1820/

Centers for Disease Control and Prevention. (1996, November 08). "Outbreak of *Escherichia coli* O157:H7 Infections Associated with Drinking Unpasteurized Commercial Apple Juice British Columbia, California, Colorado, and Washington, October 1996." MMWR 45(44); 975. Retrieved from https://www.cdc.gov/mmwr/preview/mmwrhtml/00044358.htm

Centers for Disease Control and Prevention. (2007, June 1). "Multistate Outbreak of Salmonella Serotype Tennessee Infections Associated with Peanut Butter — United States, 2006–2007." MMWR 56(21); 521–524. Retrieved from https://www.cdc.gov/mmwr/preview/mmwrhtml/mm5621a1.htm

Centers for Disease Control and Prevention. (2009, May 11). "Multistate Outbreak of *Salmonella typhimurium* Infections Linked to Peanut Butter, 2008-2009 (Final Update)." Retrieved from https://www.cdc.gov/salmonella/2009/peanut-butter-2008-2009.html

Centers for Disease Control and Prevention. (2012, August 27). "Multistate Outbreak of Listeriosis Linked to Whole Cantaloupes from Jensen Farms, Colorado (Final Update)." Retrieved from https://www.cdc.gov/listeria/outbreaks/cantaloupes-jensen-farms/index.html

Centers for Disease Control and Prevention. (2014, January 8). Estimates of Foodborne Illness in the United States. Retrieved from https://www.cdc.gov/foodborneburden/index.html#

Clayton, E. G. (1908). Arthur Hill Hassall: Physician and Sanitary Reformer. A Short History of his Work in Public Hygiene, and of the Movement Against The Adulteration of Food and Drugs. London: Bailliere, Tindall, and Cox. Retrieved from https://archive.org/details/arthurhillhassa00claygoog/page/n9

Coley, B. (2020, January). "Chipotle Fined $1.37M for Violating Child Labor Laws." QSR Magazine. Retrieved from https://www.qsrmagazine.com/employee-management/chipotle-fined-137m-violating-child-labor-laws

Coley, N. (2005, March 1). "The Fight against Food Adulteration." *Education in Chemistry*. 52(2). Royal Society of Chemistry. Retrieved from https://eic.rsc.org/feature/the-fight-against-food-adulteration/2020253.article

Department of Justice. (2009, June 8). "United States Attorneys Annual Statistical Report, Fiscal Year 1998." Retrieved from https://www.justice.gov/sites/default/files/usao/legacy/2009/06/08/98statrpt.pdf

Department of Justice. (2015, May 20). "ConAgra Subsidiary Agrees to Enter Guilty Plea in Connection with 2006 through 2007 Outbreak of Salmonella Poisoning Related to Peanut Butter." Press Release. Office of Public Affairs. Retrieved from https://www.justice.gov/opa/pr/conagra-subsidiary-agrees-enter-guilty-plea-connection-2006-through-2007-outbreak-salmonella

Department of Justice. (2020a, April 21). "Chipotle Mexican Grill Agrees to Pay $25 Million Fine to Resolve Charges Stemming from More Than 1,100 Cases of Foodborne Illness." Press Release. Retrieved from https://www.justice.gov/usao-cdca/pr/chipotle-mexican-grill-agrees-pay-25-million-fine-resolve-charges-stemming-more-1100

Department of Justice. (2020b, May 1). "Blue Bell Creameries Agrees to Plead Guilty and Pay $19.35 Million for Ice Cream *Listeria* Contamination – Former Company President Charged." Press Release. Retrieved from https://www.justice.gov/opa/pr/blue-bell-creameries-agrees-plead-guilty-and-pay-1935-million-ice-cream-listeria

Detwiler, D. (2015, September 23). "Bearing Witness to Justice at the PCA Sentencing." *Food Safety News*. Retrieved from http://www.foodsafetynews.com/2015/09/bearing-witness-to-justice-at-the-pca-sentencing/#.VgKDMN9Viko

Detwiler, D. (2016, October 24). "Death Should Not Be on the Menu." *EC Nutrition* 5.3(2016): 1148–1149. Retrieved from https://www.ecronicon.com/ecnu/pdf/ECNU-05-0000158.pdf

Detwiler, D. (2020). Food Safety: Past, Present, and Predictions. Cambridge, Elsevier Academic Press.

Dunlap, R.W. (1925, June 13). "Letter of USDA Acting Secretary R.W. Dunlap to President Calvin Coolidge." Archival Copy. Retrieved from https://6sd6hj41ya-flywheel.netdna-ssl.com/images/pdfs/109_Dunlapletter.pdf

Filby, Frederick A. (1934). A History of Food Adulteration and Analysis. London: Allen & Unwin.

Flynn, D. (2018, November 15). "Supreme Court won't review Michael Parnell's case related to deadly outbreak." *Food Safety News.* Retrieved from https://www.foodsafetynews.com/2018/11/supreme-court-wont-review-michael-parnells-case-related-to-deadly-outbreak/

House of Commons. (1872, March 6). Report from the Select Committee on Adulteration of Food etc. 1856, 379, viii, 1, p iv; Parl Deb HC vol 209. (HC 1872). London: The Stationery Office.

Howman, M. (1901). "The Sale of Food and Drugs Acts, 1875 to 1899 with Notes: Decided Cases in England and Scotland, and Appendix Containing Forms, Etc." Edinburgh: William Green and Sons.

Huang, Y. (2014, July 16). The 2008 Milk Scandal Revisited. Forbes Magazine. Retrieved from http://www.forbes.com/sites/yanzhonghuang/2014/07/16/the-2008-milk-scandal-revisited/#41a6bb204428

Jamrisko, M. (2015, April 14). "Americans' Spending on Dining Out Just Overtook Grocery Sales for the First Time Ever: Could be another reason to blame or credit millennials." *Bloomberg News.* Retrieved from https://www.bloomberg.com/news/articles/2015-04-14/americans-spending-on-dining-out-just-overtook-grocery-sales-for-the-first-time-ever

Kuhns, A., and Saksena, M. (2017, December). "Food Purchase Decisions of Millennial Households Compared to Other Generations." U.S. Department of Agriculture, Economic Research Service Economic Information Bulletin Number 186. Retrieved from https://www.ers.usda.gov/webdocs/publications/86401/eib-186.pdf?v=43097

Lely, J. M. (1894). Statutes of Practical Utility: Arranged in alphabetical and chronological order with notes and indexes: Being the 5th edition of Chittys Statutes (Vol. IV). London: Sweet & Maxwell.

Lynch, M, Painter, J., Woodruff, R., and Braden, C. (2006, November 10). Surveillance for Foodborne-Disease Outbreaks — United States, 1998–2002. *Morbidity and Mortality Weekly Report* 55: SS-10, p.2. Retrieved from http://www.cdc.gov/mmwr/PDF/ss/ss5510.pdf

Macartney, J (2008, September 22) "China Baby Milk Scandal Spreads as Sick Toll Rises to 13,000." *The Times.* Retrieved from https://www.thetimes.co.uk/article/china-baby-milk-scandal-spreads-as-sick-toll-rises-to-13000-jlxdmrsk9qd

Marler, B. (2020, April 21). "Chipotle Mexican Grill Agrees to Pay $25 Million Fine and Enter a Deferred Prosecution Agreement to Resolve Charges Related to Foodborne Illness Outbreaks." *Food Safety News.* Retrieved from https://www.foodsafetynews.com/2015/12/chipotles-first-norovirus-outbreak-in-california-was-larger-than-boston-colleges/

Mead, P., Slutsker, L., Dietz, V., McCaig, L., Bresee, J., Shapiro, C., et al. (1999, October). Food-Related Illness and Death in the United States. Centers for Disease Control and Prevention. Retrieved from http://wwwnc.cdc.gov/eid/article/5/5/99-0502

Mews, J., Gordon, W. E., and Spencer, A. J. (Eds.). (1896). The Law Journal Reports for the Year 1896. Volume 65, Page 19. London: Stevens and Sons.

Milazzo, D. (2015, April 23). Report from Preventive Controls Team for FSMA Implementation. The FDA Food Safety Modernization Act: Focus on Implementation for Prevention-Oriented Food Safety Standards Public Meeting. Presentation for FDA, Washington, DC.

Morling, A. (2017). 2017 Food Defense Conference. Minneapolis, MN: National Center for Food Protection and Defense.

National Center for Food Protection and Defense. (2011, April 30). Backgrounder: Defining the Public Health Threat of Food Fraud. St. Paul, Minneapolis. Retrieved from http://foodfraud.msu.edu/wp-content/uploads/2014/07/food-fraud-ffg-backgrounder-v11-Final.pdf

Ostroff, Stephen. (2016, November). "FDA's Take on Criminal Liability." Plenary Session and Town Meeting conducted at the Food Safety Consortium, Schaumburg, IL.

Powell, D. (2018, December 2). "Your Vomit and Diarrhea is Our Bread and Butter: Portland's Outbreak Museum." *Barf Blog*. Retrieved from https://www.barfblog.com/tags/bill-keene/

Ryan, J. and Glarum, J. (2008). Biosecurity and Bioterrorism: Containing and Preventing Biological Threats. Burlington, MA: Elsevier.

Schulz, D. and Kopke, U. (1997). The Quality Index: A Holistic Approach to Describe Quality of Food. Retrieved from http://orgprints.org/2519/1/SCHULZ_KOEPKE_1997a_e.pdf

Sinclair, U. (1906, March 10). "Letter from Upton Sinclair to President Theodore Roosevelt, 03/10/1906 (National Archives Identifier: 301981); Letters Received, 1893 - 1906; Records of the Office of the Secretary of Agriculture, 1839 - 1981; Record Group 16; National Archives. Retrieved from https://www.archives.gov/historical-docs/todays-doc/?dod-date=310

Spink, J. and Moyer, D.C. (2011). Defining the Public Health Threat of Food Fraud. *Journal of Food Science* 76(9): R157–R162. Retrieved from http://foodfraud.msu.edu/wp-content/uploads/2013/03/Article-Understanding-and-Combating-Food-Fraud-FT-Food-Technology-2013-01-b.pdf

The Franco-American Food Company (1906, June 8). "An Open Letter to President Roosevelt and the American Nation." *The New York Times*. Microfilm collection, Western Washington University.

Theunissen, M. (2016, December 18). "New Kiwi Software Intercepts Fake Infant Formula." *New Zealand Herald*. Retrieved from http://www.nzherald.co.nz/business/news/article.cfm?c_id=3&objectid=11769310

U.S. Department of Agriculture. (1887). Technical Bulletin 19330 – Number 13: Foods and Food Adulterants. Retrieved from https://archive.org/details/foodsfoodadulter13unit/page/n10

U.S. Department of Agriculture. (2018, February 21). "FSIS History." Retrieved from https://www.fsis.usda.gov/wps/portal/informational/aboutfsis/history

U.S. Department of Justice. (2013A, February 15). "United States v. Stewart Parnell, Michael Parnell, Samuel Lightsey, and Mary Wilkerson, Defendants." CASE NO.1:13-CR1-2 in the United States District Court for the Middle District of Georgia, Albany Division. Retrieved from https://www.justice.gov/iso/opa/resources/61201322111426350488.pdf

U.S. Department of Justice. (2013B, October 22). "Eric and Ryan Jensen Plead Guilty to All Counts of Introducing Tainted Cantaloupe into Interstate Commerce." [Press Release]. Retrieved from https://www.justice.gov/usao-co/pr/eric-and-ryan-jensen-plead-guilty-all-counts-introducing-tainted-cantaloupe-interstate

U.S. Department of Justice. (2015A, April 13). "Quality Egg, Company Owner and Top Executive Sentenced in Connection with Distribution of Adulterated Eggs." [Press Release]. Retrieved from https://www.justice.gov/opa/pr/quality-egg-company-owner-and-top-executive-sentenced-connection-distribution-adulterated

U.S. Department of Justice. (2015B, September 21). "Former Peanut Company President Receives Largest Criminal Sentence in Food Safety Case; Two Others also Sentenced for their Roles in Salmonella-Tainted Peanut Product Outbreak." [Press Release]. Retrieved from https://www.justice.gov/opa/pr/former-peanut-company-president-receives-largest-criminal-sentence-food-safety-case-two

U.S. Food and Drug Administration. (2018, September 12). "FSMA Final Rule for Mitigation Strategies to Protect Food Against Intentional Adulteration." Retrieved from https://www.fda.gov/food/food-safety-modernization-act-fsma/fsma-final-rule-mitigation-strategies-protect-food-against-intentional-adulteration

United Nations Food and Agriculture Organization. (2009). "Declaration of the World Food Summit on Food Security." Retrieved from http://www.fao.org/fileadmin/templates/wsfs/Summit/Docs/Final_Declaration/WSFS09_Declaration.pdf

United Nations. (2015). "Transforming Our World: the 2030 Agenda for Sustainable Development." Retrieved from https://sustainabledevelopment.un.org/post2015/transformingourworld

Wedderburn, A. (1890). U.S. Department of Agriculture Technical Bulletin 25: "A Popular Treatise on the Extent and Character of Food Adulterations." Retrieved from https://archive.org/details/populartreatiseo25wedd/page/n2

Wiley, H. (1925, September). "Letter to President Coolidge: Enforcement of the Food Law." *Good Housekeeping Magazine*. Archived copy. Retrieved from https://6sd6hj41ya-flywheel.netdna-ssl.com/images/pdfs/53_Letters%20to%20President.pdf

Wilson, B. (2005, February 27). "Food Scares are One of the Great British Traditions." *The Telegraph*. Retrieved from https://www.telegraph.co.uk/news/uknews/1484479/Food-scares-are-one-of-the-great-British-traditions.html

Wittenberger, K., and Dohlman, E. (2010, February). "Peanut Outlook: Impacts of the 2008-09 Foodborne Illness Outbreak Linked to Salmonella in Peanuts." USDA Economic Research Service Report OCS-10a-01. Retrieved from https://www.ers.usda.gov/webdocs/publications/37835/8684_ocs10a01_1_.pdf

World Health Organization. (2003). Terrorist Threats to Food: Guidance for Establishing and Strengthening Prevention and Response Systems. Retrieved from https://apps.who.int/iris/handle/10665/42619

World Health Organization. (2015, March 12). WHO Estimates of the Global Burden of Foodborne Diseases. Geneva, Switzerland. Retrieved from http://apps.who.int/iris/bitstream/10665/199350/1/9789241565165_eng.pdf

chapter two

Codex Alimentarius and transparency of the global food safety system

Alexey Petrenko and Victor Tutelyan

Contents

EDITORS' NOTE: GLOBAL FOOD SAFETY REGULATION, TRADE, AND ANTIDUMPING

Chapter 2 leads off with experts Petrenko and Tutelyan highlighting the links of the Codex Alimentarius, trade agreements under the World Trade Organization's (WTO) food safety purview, and fair trade to food system transparency. First, the authors characterize the Codex Alimentarius Commission's (CAC) decision-making process to correlate how international trading partners are working to protect the health and safety of all consumers. Facilitated by the WTO and Food and Agriculture Organization (FAO), CAC subcommittees and their budgetary needs are introduced to demonstrate how transparency within globally recognized trade agreements are created and amended.

Further, this chapter discusses antidumping. An antidumping law is "[a] statute designed to protect domestic companies by preventing the sale of foreign goods at less than fair value, as defined in the statute (for example, at a price below that of the domestic market)."[1] Conversely, dumping is defined as:

1. The act of selling a large quantity of goods at less than fair value.
2. Selling goods abroad at less than the market price at home.[2]

To illustrate,

> [d]umping has traditionally been defined as the type of price discrimination between national markets, in which a producer sells at a lower price abroad than in his home market (price dumping …). It is often considered unfair that a producer, who benefits from protection in his home market and therefore can charge high prices there, subsequently uses the artificially high profits generated on the protected home market to subsidize low-priced export sales…. Over time, … so-called cost dumping, has been treated as actionable. Under this concept, anti-dumping duties may effectively be imposed on producers that sell below full cost of production in an export market.[3]

An example of an antidumping law in the context of food trade are importers receiving a bulk shipment of grain or frozen seafood at a price below the exporter's domestic market rate. In the United States, the U.S. Department of Commerce's International Trade Administration enforces antidumping laws to protect American businesses from these deceptive practices.

One of the potential reactions to dumping might be countervailing measures, also discussed in Chapter 2. A countervailing measure is an "[a]ction taken by the importing country, usually in the

---

[1] ANTIDUMPING LAW, Black's Law Dictionary (11th ed. 2019).
[2] DUMPING, Black's Law Dictionary (11th ed. 2019).
[3] Id. citing Edwin Vermulst, "Anti-Dumping," in 1 *The Max Planck Encyclopedia of Public International Law* 436, 437 (Rüdiger Wolfrum ed., 2012).

form of increased duties to offset subsidies given to producers or exporters in the exporting country."[4]

The WTO is an international body overseeing treaties between countries that trade, including in food and other agricultural products. Specifically, the WTO is the "only global international organization dealing with the rules of trade between nations. At its heart are the WTO agreements, negotiated and signed by the bulk of the world's trading nations and ratified in their parliaments. The goal is to ensure that trade flows as smoothly, predictably and freely as possible."[5] Correspondingly, WTO member countries, also called "signatories," enter into treaties that set up rules for a level playing field in trade. Several of these treaties affect food trade, such as the General Agreement on Tariffs and Trade (GATT), for instance.

When countries that are WTO signatories attempt to protect their internal markets from dumping, they may be "safeguarding," or taking safeguard measures, which means that a "WTO member may take a 'safeguard' action (i.e., restrict imports of a product temporarily) to protect a specific domestic industry from an increase in imports of any product which is causing, or which is threatening to cause, serious injury to the industry." In fact, under GATT Article XIX, the provision on Emergency Action on Imports of Particular Products "[s]afeguard measures were always available ... [but] infrequently used, and some governments preferred to protect their industries through 'grey area' measures ('voluntary' export restraint arrangements on products such as cars, steel and semiconductors)." Additionally, the WTO Safeguards Agreement prohibits "grey area" measures and setting time limits ('sunset clause') on all safeguard actions.[6] This means that arbitrary methods to resolve trade disputes are not sufficient in ensuring fair trade. Instead, these disputes must follow established rules and practices set forth by the WTO to provide increased visibility in safeguard investigations and actions for governments and citizens.

Other agreements under the WTO's purview include the Technical Barriers to Trade (TBT) Agreement, which:

---

[4] WTO Glossary, counterveiling measures, https://www.wto.org/english/thewto_e/glossary_e/countervailing_measures_e.htm
[5] WTO, https://www.wto.org/english/thewto_e/thewto_e.htm
[6] Safeguard measures, WTO glossary, https://www.wto.org/english/tratop_e/safeg_e/safeg_e.htm. Statistics on safeguard measures can be tracked at https://www.wto.org/english/tratop_e/safeg_e/safeg_e.htm#statistics

aims to ensure that technical regulations, standards, and conformity assessment procedures are non-discriminatory and do not create unnecessary obstacles to trade. At the same time, it recognises WTO members' right to implement measures to achieve legitimate policy objectives, such as the protection of human health and safety, or protection of the environment. The TBT Agreement strongly encourages members to base their measures on international standards as a means to facilitate trade. Through its transparency provisions, it also aims to create a predictable trading environment.[7]

The creation of such a predictable trading environment means that barriers to trade must be minimized and standards harmonized. How does this work in the context of food and agricultural trade and where does transparency come into play?

According to the WTO's own account, "[t]echnical regulations and standards are important, but they vary from country to country. Having too many different standards makes life difficult for producers and exporters. If the standards are set arbitrarily, they could be used as an excuse for protectionism."[8] This means that "[t]he protection of domestic businesses and industries against foreign competition by imposing high tariffs and restricting imports"[9] impairs the free trade that the WTO seeks to protect. Thus, discrepancies in food safety standards, for instance, "can become obstacles to trade. But they are also necessary for a range of reasons, from environmental protection, safety, national security to consumer information. And they can help trade. Therefore, the same basic question arises again: how to ensure that standards are genuinely useful, and not arbitrary or an excuse for protectionism?"[10] This is what the TBT agreement regulates by ensuring that "regulations, standards, testing and certification procedures do not create unnecessary obstacles"[11] to trade.

---

[7] https://www.wto.org/english/tratop_e/tbt_e/tbt_e.htm
[8] https://www.wto.org/english/thewto_e/whatis_e/tif_e/agrm4_e.htm#TRS
[9] PROTECTIONISM, Black's Law Dictionary (11th ed. 2019).
[10] Id.
[11] Id.

Another important agreement alongside the TBT is the Application of Sanitary and Phytosanitary Measures (SPS Agreement). The SPS Agreement, in turn, strives "to improve the human health, animal health and phytosanitary situation in all Members" to the WTO by:

> further the use of harmonized sanitary and phytosanitary measures between Members, on the basis of international standards, guidelines and recommendations developed by the relevant international organizations, including the Codex Alimentarius Commission, the International Office of Epizootics, and the relevant international and regional organizations operating within the framework of the International Plant Protection Convention, without requiring Members to change their appropriate level of protection of human, animal or plant life or health.[12]

In the context of food trade, these principles of economics – free trade, prevention of protectionism, and minimization of trade obstacles – overlap with the sheer requirements to ensure transparency in trade, i.e., information flow in food trade, thus Food System Transparency. The following chapter identifies the challenges to maintain transparency and food safety in the complicated realm of global food trade. The explanations in this chapter cover important topics from the role of the Codex Alimentarius to fair trade from the perspective of food system transparency.

However, principles of economics and free trade are not the only ones at the heart of increasing food system transparency on an international level. Sustainable development is a priority of prudent regulation and the United Nation's FAO works with the WTO toward shared goals:

> Food standards and trade go hand in hand in ensuring safe, nutritious and sufficient food for a growing world population. The Sustainable Development Goals (SDGs) acknowledge the

[12] https://www.wto.org/english/tratop_e/sps_e/spsagr_e.htm

role that trade can play in promoting sustainable development. Together, FAO and the
WTO and their international system of food
standards and trade contribute to achieving SDG 2 on hunger, food security, nutrition
and sustainable agriculture; SDG 3 on healthy
lives and wellbeing; SDG 8 on economic
growth, employment and work; and SDG 17
on strengthening global partnerships for sustainable development.

...

food is difficult to imagine without standards. Food standards give confidence to consumers in the safety, quality and authenticity
of what they eat. By setting down a common
understanding on different aspects of food
for consumers, producers and governments,
standards enable trade to take place. If every
government applies different food standards,
trade is more costly, and it is more difficult to
ensure that food is safe and meets consumers'
expectations.[13]

As such, it is more important than ever to ensure that food standards are scientifically sound, transparent, and flexible. The Codex
Alimentarius, as the following chapter explains, provides a strong
framework to preserve free and unhindered trade while also stream-
lining food standards and the information required to maintain
them. In short, the following chapter shows the key position that the
Codex takes for *Food System Transparency*.

[13] https://www.wto.org/english/res_e/booksp_e/tradefoodfao17_e.pdf

Challenges in the international food supply

Since World War II, the deadliest military conflict in history, saw 3 percent of humankind disappearing, the world's population has been tri-
pling through the period of relative peace and stability (United Nations,
2019). Poverty level, measured as a share of people living on $1.90 per
day, has dropped from 42 percent of the global population in the 1960s to
10 percent in late 2010s (World Bank, 2020). This decline is primarily due

to considerable growth in the global agricultural output that, in the same period, increased fourfold reaching $3.4 trillion (FAO, 2020).

In the second half of the 20th century, innovations in food production, coupled with the networking of the regional logistics routes, triggered unprecedented globalization of the food supply. Multiplying increases in production yields resulted in the accumulation of the food stock in food-rich regions, which, with the growing volume of the international trade, could be shared with food-deficient parts of the world.

However, access to food as a basic human need remains a global challenge. Eight hundred million people go hungry every day, and malnutrition affects one in three people. In the list of the seventeen 2030 Sustainable Goals adopted by the United Nations' General Assembly in 2015, ending poverty and hunger, as well as promoting sustainable agriculture, are top priorities (United Nations, 2015).

At the same time, the globalization of the food supply makes it ever so difficult to ensure safety of foods traveling long distances. Risk levels of life-threatening contamination affecting large population groups grew at all links of the supply network. The Aberdeen typhoid outbreak in 1964[14] (Smith, 2007) and the spread of mycotoxins contamination (Peraica et al., 1999) in Africa became the two alarming examples of new threats associated with the globalized food supply.

Academic experts believe that, despite its size and outreach, the global food system is unable to provide access to safe and quality food for all population groups across the world (InterAcademy Partnership, 2018). In general, the sustainability of commodity production and trade depends on three groups of factors: the scale, distribution of responsibility, and system transparency (Gardner et al., 2019). Defined as "the extent to which all the network's stakeholders have a shared understanding of, and access to, product and process related information that they request, without loss, noise, delay and distortion" (Beulens et al., 2005), transparency on the global food supply is proportional to the degree of shared information and compliance with commonly binding food safety and quality criteria.

Totally, 73.2 percent of the world's agricultural production is concentrated in the fifteen countries listed in Table 2.1.[15] These countries produce, on average, $470 worth of food per capita, while the rest of the world's food output is close to $310 worth of food per capita. Not accidentally, the fifteen countries are also responsible for 80 percent of global food exports and

---

[14] Several cases of typhoid were reported in Aberdeen after infected South American beef contaminated a slicing machine in a retail shop.
[15] The European Union (EU) is treated here as a single economic entity with the output of EU countries totalled to a single value.

Table 2.1 World's largest food producers and value of their agricultural
production, in constant dollars, 2018

Country	Population, mil	Value-added agri production, billion USD	Per capita, USD	Year in CAC
China	1,393	770.336	550	1984
India	1,353	375.033	270	1964
European Union	531.2	272.248	510	2003
United States	327.2	178.321	550	1963
Indonesia	267.7	143.178	520	1971
Nigeria	195.9	116.729	600	1969
Brazil	209.5	114.363	520	1968
Turkey	81.1	86.356	1,100	1963
Russia	144.5	64.665	460	1993
Japan	126.5	53.878	420	1963
Mexico	126.2	41.373	320	1969
Argentina	44.5	31.171	690	1963
Canada	37.1	27.499	730	1963
Australia	25.0	26.444	1,000	1963
New Zealand	4.9	16.097	3,300	1963
Total for 15	4,867.3	2,317.7	470	
Rest of World	2,727	847	310	

76 percent of food imports (FAO, 2020). Essentially, they have long become
major global/regional knots in the worldwide food supply network.

With so many powerful players, the food market is a place of fierce
competition, which induces distortions in information flow and provokes
nationalistic trade policies that not only reduce supply chain transparency, but also harm interests of global consumers.

From the early days of the East India Company and *Vereenigde Oost-
Indische Compagnie* (VOC) (Robertson and Funnell, 2012), established by
the English and Dutch to control spice trade routes in the early 17th century, up until the most recent U.S.-China tariff dispute, protectionism
has been a popular tool commonly used to fence local agricultural markets from the international competition. Regardless of the World Trade
Organization's (WTO) long-standing effort to reduce trade tariffs, countries readily introduce temporary protection measures: antidumping,
safeguarding, and countervailing measures, which have been found to be
inversely proportional to the WTO tariff reductions goals (Kuenzel, 2020).

The international food trade has also been subject to numerous technical barriers to trade (TBT) and sanitary and phytosanitary (SPS) measures,
which are claimed to be required for health and safety reasons, but, in

reality, serve as an effective protectionism device. A recent study (Melo et al., 2014) evaluated the impact of sanitary, phytosanitary, and quality-related regulations and standards on international trade, concluding that the more stringent the regulatory measures, the more negative effect they have on export volumes. Remarkably, the negative effect is more pronounced for TBTs and SPSs introduced for exports by developed economies.

There is a thin line in policy management between the need to protect human health and the need to protect local manufacturers from increasing competition abroad. Decisions to introduce a TBT or an SPS measure often lack justification, which should be clearly communicated to all stakeholders. This is an unambiguous sign that the food supply transparency has been compromised.

The WTO Agreement on Technical Barriers to Trade (World Trade Organization, 1995a) and the Agreement on Sanitary and Phytosanitary Measures recognize (World Trade Organization, 1995b) the right of a member state to impose technical regulations, conformity assessment procedures, and SPS measures to achieve legitimate policy objectives, such as the protection of human health and safety or protection of the environment. At the same time, both agreements aim to ensure that these measures remain nondiscriminatory and do not create unnecessary obstacles to trade goods.

The TBT and SPS Agreements attempt to address the issue of transparency strongly encouraging members to base their measures on international standards.

Codex Alimentarius and its global reputation

When it comes to food, the only choice that countries around the world have for internationally recognized standards is Codex Alimentarius, the food code that is developed to ensure fair trade practices and safety of consumers globally.

The WTO SPS agreement makes direct reference to the Codex Alimentarius as the set of international standards to which sanitary or phytosanitary measures have to conform to be deemed necessary to protect human, animal, or plant life or health, and presumed to be consistent with the relevant provisions of the agreement:

> International standards, guidelines and recommendations for food safety [are] the standards, guidelines and recommendations established by the Codex Alimentarius Commission relating to food additives, veterinary drug and pesticide residues, contaminants, methods of analysis and sampling, and codes and guidelines of hygienic practice.

The idea of Codex Alimentarius originated from Codex Alimentarius Austriacus, a set of standards and guidelines adopted in 1911 in the Austria-Hungarian Empire, whose enormous and widely spread army suffered from frequent food poisonings and low-quality food supply (Vojir, Schübl, and Elmadfa, 2012). In 1954, Austria also led the process of creating Codex Alimentarius Europaeus – a regional food code to which major food-producing countries agreed to adhere in post-war Europe.

However, this regional effort was firmly resisted by the Food and Agriculture Organization (FAO) of the United Nations and the World Health Organization (WHO), both arguing for a harmonized global regulatory approach to food safety and trade, and for establishing a global food standardization body.

In 1961, the 11th session of FAO agreed to launch a Codex Alimentarius Commission (CAC), and adopted its statutes (FAO, 1961) with membership open to all FAO, WHO, and UN members. The first CAC session was held in Rome in 1963 on the recommendation of the Joint FAO/WHO Conference on Food Standards and was attended by twenty-seven member countries (Joint FAO/WHO Codex Alimentarius Commission, 1963).

One of the major issues of the time was (and remains) the lack of standardized analytical methods used for testing contaminants, food additives, and essential nutrients, e.g., vitamins and minerals in commercial foods. The Joint FAO/WHO Expert Committee on Nutrition recommended (FAO and WHO, 1950) that FAO, in partnership with national organizations, should generate work on analytical methods for measuring vitamin and mineral contents in the whole variety of foods. The committee also emphasized that food regulations in different countries are often conflicting and contradictory:

> …legislation governing preservation, nomenclature and acceptable food standards often varies widely from country to country. New legislation not based on scientific knowledge is often introduced, and little account may be taken of nutritional principles in formulating regulations.

From the very beginning, CAC offered all members an equal opportunity to participate in the development of international standards, adherence to which aims to protect the health and safety of consumers. Major food-rich countries sought to secure their leadership in the new organization recognizing an opportunity to create a favorable regulatory environment for unlocking new markets and expanding international trade. With its state membership growing steadily and reaching 189 member countries in 2018, CAC has also become a popular platform for industry lobbyists. By 2020, CAC has registered 236 observers including 163 international nongovernmental organizations that represent interests of all food industrial sectors, from winemaking to chewing gum producers.

Today, CAC has been recognized by all stakeholders as the only international forum which provides the global community with a comprehensive collection of food safety standards and methodologies. Its science-based approach, coupled with a consensus-based decision-making process, has won CAC a solid reputation as an indisputable authority in food standardization. Though Codex standards and related texts are voluntary, they need to be translated into national legislation in order to be enforceable. Dozens of countries around the world blindly duplicate Codex documents in their national regulations without any review or discussion.

Since its foundation in 1963, CAC has produced 224 standards, 79 guidelines, and 54 codes of practice. Most of them have already found their way into regional and national regulations, while the General Standard of Food Additives (GSFA) CODEX STAN 192-1995 (Joint FAO/WHO Codex Alimentarius Commission, 2019a) and General Standard for Contaminants and Toxins in Food and Feed CODEX STAN 193-1995 have been recognized as the only reference points for food additives use and contaminants maximum levels (MLs) around the world.

The United Nations has acknowledged the CAC for their role in addressing food security. In particular, Article 39 of Resolution of the UN General Assembly A/RES/39/248 of 16 April 1985 reads:

> When formulating national policies and plans with regard to food, Governments should take into account the need of all consumers for food security and should support and, as far as possible, adopt standards from the FAO/WHO Codex Alimentarius.

CAC lists transparency as its core value (Joint FAO/WHO Codex Alimentarius Commission, 2019b), which is translated into (FAO, 2018):

- Transparency of scientific advice provided by FAO/WHO expert bodies. This is critical for the risk analysis (discussed later in this chapter).
- Transparency of Codex information with all documents available to all stakeholders and general public.
- Transparency of its structure and internal rules and procedures.

Challenges to governance and structure

The WTO's SPS Agreement explicitly permits governments to choose not to use CAC standards and recommendations. However, if the national requirement results in a higher restriction of international trade, a country will be asked to provide scientific justification demonstrating that the relevant CAC standard would not result in the level of health protection the country considered appropriate.

Thus, Codex Alimentarius serves as an independent benchmark in assessing if there is discrimination against imported foods, which is prohibited by the WTO. In other words, while Codex recommendations may or may not be accepted as such, the SPS agreement helps them play a fundamental role in regulating international food trade. Furthermore, WTO members are required to submit scientific justification for import restrictions based on national regulations that are not harmonized with Codex Alimentarius.

Subsequently, parent organizations – FAO and WHO – gave the CAC the following mandate (Joint FAO/WHO Codex Alimentarius Commission, 2018):

- Protecting the health of consumers and ensuring fair practices in the food trade.
- Coordinating all food standards work undertaken by international governmental and nongovernmental organizations.
- Determining priorities and initiating and guiding the preparation of draft documents.
- Finalizing standards and publishing them in a Codex Alimentarius either as regional or worldwide standards.
- Amending published standards, after appropriate survey in the light of developments.

To serve these purposes, CAC must demonstrate the highest degree of transparency in its structure and governance, as well as follow an immaculate decision-making process.

The CAC has a bulky structure. Every year, the CAC annual meeting elects the chairperson and three vice-chairpersons, who can be re-elected twice, provided that by the end of their second term of office, they have not served for more than two years. The chairs can preside at meetings; however, they have no real executive power to exercise.

The CAC also appoints regional coordinators who administer the work of Codex Coordinating Committees. These committees are dedicated regional committees for Africa, Asia, Europe, Latin America and the Caribbean, Near East, North America, and the Southwest Pacific.

At the top of the CAC structure sits the Executive Committee, which acts as the executive body between CAC annual meetings. The Executive Committee can make proposals regarding general orientation, strategic planning, and programming of the CAC work. They can also study special problems and assist in the management of the standards development program, namely by conducting a critical review of proposals to undertake new work and monitoring the progress of documents approved for development.

Codex subsidiary bodies – the committees and task forces hosted or cohosted by selected member countries – take responsibility for developing, editing, and revising all Codex drafted documents. Subsidiary bodies are as follows:

- **General subject committees** (horizontal committees): as the name suggests, these committees deal with matters applicable to all foods and draft horizontal (non-product specific) standards. Ten active committees are:
  - Committee on General Principles, Committee on Food Additives;
  - Committee on Contaminants in Foods;
  - Committee on Food Hygiene, Committee on Food Labeling;
  - Committee on Methods of Analysis and Sampling;
  - Committee on Food Import and Export Inspection and Certification Systems;
  - Committee on Pesticide Residues, Committee on Residues of Veterinary Drugs in Foods; and
  - Committee on Nutrition and Foods for Special Dietary Uses.
- **Commodity committees** (vertical committees): these committees specialize in issues attached to specific food categories. The following committees are currently in operation:
  - Codex Committee on Spices and Culinary Herbs;
  - Committee on Fats and Oils;
  - Committee on Fresh Fruits and Vegetables; and
  - Committee on Processed Fruits and Vegetables.
- *Ad hoc* **intergovernmental task forces**: task forces are formed to work on a specific issue, and they operate for a fixed period. There is only one task force currently active, hosted by The Republic of Korea, working on antimicrobial resistance (AMR).

While the role of subsidiary bodies is recognized by members and well understood by the public, the role of the Executive Committee is not always clear. They are said to have an advisory role offering chairpersons and regional coordinators an opportunity to formulate matters which the FAO and WHO want the CAC to consider, as well as to review the CAC budget. Nevertheless, all CAC decisions are made or sanctioned by the annual meeting so, like chair and vice-chairs, the Executive Committee has no actionable executive power.

The CAC and the Executive Committee are administered by the Codex Secretariat, an administrative body employed at the expense of CAC to run all administrative matters, i.e., meeting schedules, agendas, reporting, and minute-taking. The Secretariat also serves as an expert authority to advise members on Codex procedures and rules. This advice often comes during major disagreements during in-session meetings with subsidiary bodies, helping chairs to progress through the meeting agendas.

CAC is funded by WHO and FAO (roughly at 30:70 proportion, respectively); however, it is a remarkable feature of the Codex structure that subsidiary bodies (except coordinating (regional) committees) are funded by hosting member countries. For example, the German government would bear all costs for chairing the Codex Committee on Nutrition and Foods for Special Dietary Purpose. These costs include not only staff and administrative matters, but also hosting annual committee meetings.

While host countries declare their efforts to ensure transparency of committee discussions and decisions, they unavoidably exert a certain degree of influence over some work by selecting documents and proposals, which are prioritized by their committees and passed to the CAC annual meetings for adoption.

Thus, CAC's executive power has shifted from formally appointed executive authorities – chairpersons and the Executive Committee – to committee chairs and Codex Secretariat. There is no coincidence that all committee chairs are currently occupied by major food-producing and trading countries listed in Table 2.1,[16] namely, the United States, China, Australia, Canada, India, Mexico, and four European Union (EU) members – Germany, the Netherlands, France, and Hungary.

While some Codex member countries find the status quo beneficial for their resource management, critics use this fact to demonstrate that CAC serves the interests of major food-producing nations helping them to control the global food supply network (Hamer, 2013). After all, the committee chairs were also among the first countries to join CAC, being very active in all subsidiary bodies.

Figure 2.1 maps the most active Codex member states together with the 2018 value of their food trade volume. The filled map shows how active the Codex member countries are with darker shade, representing high activity in Codex activities. It can be seen in the figure that the more intense the country's shade is the higher is its food trade value.

CAC seems to be aware of the issue. In the current five-year Codex Strategic Plan (Joint FAO/WHO Codex Alimentarius Commission, 2019b), one of the goals is to facilitate the participation of all Codex Members throughout the standard-setting process.

The organization also seeks to enhance work management systems and practices that support the efficient and effective achievement of all strategic plan goals. While this might sound too broad, there are certain challenges on the way to achieve these objectives.

---

[16] European countries that became Codex members in 1963 are now part of the EU.

Figure 2.1 Activity of member countries in CAC, measured by their participation in executive structures, subsidiary bodies and working groups, and value of their international food trade (import plus export in billions of dollars) in 2018.

Transparency of Codex risk analysis: Risk assessment and risk communication

Risk assessments and risk communications are major components to help Codex Alimentarius members provide greater food system transparency throughout the global supply chain. These assessments and communications require technical expertise to identify, evaluate, and ultimately communicate risk to key stakeholders, including government, industry, and consumers.

The WHO defines risk assessment as "the scientific evaluation of known or potential adverse health effects resulting from human exposure to foodborne hazards." The process consists a hazard identification, hazard characterization, exposure assessment, and risk characterization.[17] The WHO provides quantitative risk assessments and qualitative expressions of risk for a given population through numerical expressions and defining levels of uncertainty.

Risk communication serves a critical role to disseminate serious public health information to the general public. Primarily in the food system,

[17] https://www.who.int/foodsafety/micro/riskassessment/en/#:~:text=Risk%20assessment%20is%20the%20scientific,associated%20with%20a%20particular%20agent

risk communications are used to strengthen trust between food industry stakeholders and the public. They normally incorporate methods to
discuss preparedness, response, and recovery plans following a serious
public health event.

Codex Alimentarius documents must be fully and systematically documented in a transparent manner. They must be:

- **Science-based:** developing and adopting Codex documents, CAC
  must demonstrate that sound up-to-date scientific evidence was
  considered.
- **Relevant to public health and well-being:** food safety criteria and
  codes of practices form a large part of Codex work, while logistics
  and manufacturing issues that do not pose a health risk to consumers would not be included in the Codex agenda.
- **Relevant to global (regional) community:** only global or regional
  issues are considered.
- **Consensus-based:** document adoptions take place only after members reach full agreement. Though in practice, decisions in Codex
  could still go through if a small number of member countries remain
  in opposition (express their reservation).
- **Supported by the science-based four-step risk assessment:** in
  accordance with the Statements of Principle Relating to the Role
  of Food Safety Risk Assessment, firms should incorporate the four
  steps of the risk assessment, including hazard identification, hazard
  characterization, exposure assessment, and risk characterization.

CAC's Procedural Manual reads that food safety and health aspects of
Codex standards and related texts should be based on the comprehensive
risk analysis. Essentially, CAC is a risk analysis platform offering food
safety authorities of member countries risk management options based on
independent scientific advice, related to food and feed safety with respect
to public health issues (risk assessment).

The risk assessment model (Figure 2.2) includes hazard (a chemical,
microbiological, or physical contaminant) identification based on exposure assessment, i.e., assessment to what level consumers are exposed
to the hazard, and hazard analysis, i.e., defining what threat the hazard
brings to human health and well-being. Both studies require data from
member countries, including records on food consumption and toxicolog-
ical studies. At the next step, the hazard analysis and the exposure assessment are integrated to produce risk characterization and risk estimate as
low, medium, or high.

While responsibility for the risk management lies with CAC and its
subsidiary bodies (risk managers), in risk assessment, CAC relies primarily on the Joint FAO/WHO expert bodies and consultations (risk assessors).

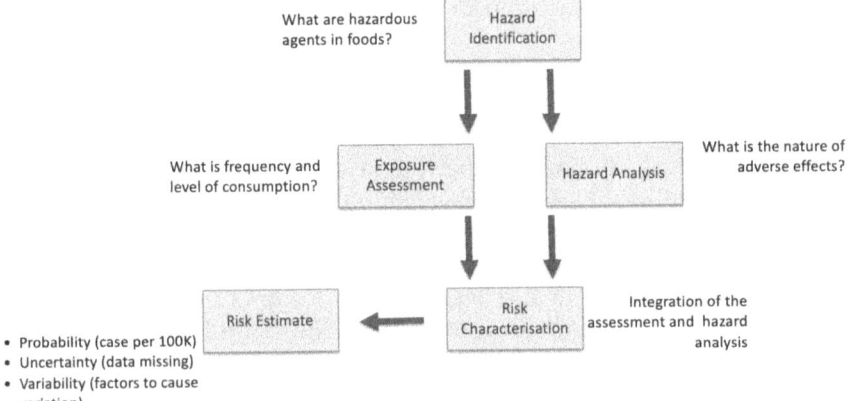

Figure 2.2 Risk assessment model used by CAC.

Risk assessment and risk management are linked by the third component of risk analysis – risk communication – the interactive exchange of information and opinions generated throughout the risk analysis process including risks, risk-related factors, and risk perception between stakeholders involved: risk assessors, risk managers, consumers, industry, and the academic community.

Risk communication needs to clearly explain risk assessment findings and the basis of risk management decisions.

At the end, efficiency of the whole process of risk analysis is measured by preparedness of authorities in Codex member countries to deal with a food safety crisis. In particular, the creation of various tools, such as templates for data gathering, situation report templates, and decision trees, as well as clear and concise reference materials for use during emergencies, can limit the number of decisions that the emergency risk managers will have to make under time constraints (FAO and WHO, 2011).

In this regard, CAC communication to the external world, including all Codex standards, guidelines, and recommendations, is the risk communication which should not only ensure a sound basis for understanding the risk management decisions offered but also foster public involvement in the process to enhance trust and confidence in the safety of the global food supply.

However, instead of taking full leadership of the process, CAC delegates essential parts of the risk communication to the Codex Contact Points in member countries, relying on them to distribute Codex work and ensure that new documents are taken into account by national authorities and consumers. The proactive communication is reduced to distribution of meeting reports and publishing of new documents on the CAC website without a further explanation providing why and how the new documents should be used.

There are hundreds of documents released every year by CAC, its subsidiary bodies, working groups, and members themselves. Most of them need to be considered in the context of lengthy discussions that take place over several years. Boundaries between risk assessments and risk management blur allowing risk management decisions to move away from risk estimates to arguments shaped by political and economic context.

As a result, the transparency of the risk analysis gets reduced. One example is a recent discussion in the Codex Committee on Food Additives (CCFA) of Note 161 used in the GSFA (Joint FAO/WHO Codex Alimentarius Commission, 2019a). The note was introduced back in 2009 by CCFA41 session, (CCFA, 2009) when the argument broke over the justification of the use of sweeteners for reduction of the calorie value of foods. Some countries insisted that they were not in the position to approve the use of sweeteners in foods other than those with reduced energy value or no added sugars. They proposed that provisions for sweeteners in the GSFA could only be approved with the note attached:

> Subject to national legislation of the importing country aimed, in particular, at consistency with Section 3.2 of the Preamble.

Section 3.2 of the GSFA preamble reads that a food additive can be used to enhance the keeping quality or stability of a food or to improve its organoleptic properties, providing that this does not change the nature, substance, or quality of the food, so as to deceive the consumer.

Countries that advocated for the use of note 161 observed that the use of sweeteners in foods, which did not bear "low-energy" or "no-sugar" claim, was misleading for consumers. Other countries saw that adopting sweetener provisions with note 161 undermined the whole purpose of CAC to produce global international standards. As a result, there was no consensus in approving the majority of sweetener (and some color) provisions that were held without approval for over ten years.

Experts not familiar with the ten-year-long discussion struggled to understand the importance of the breakthrough achieved at the fifty-first session of CCFA in 2019, which allowed immediate consensus with the note text replaced with the new edition:

> Some Codex members allow use of additives with sweetener function in all foods within this Food Category while others limit additives with sweetener function to those foods with significant energy reduction or no added sugars.

The new wording emphasized that there are regional/national differences in countries' views on how consumers may be misled. However, the discussion deviated from the CAC principles of risk analysis, as it failed to explain two aspects.

First, Joint FAO/WHO Expert Committee on Food Additives (JECFA) safety assessments conducted for all sweeteners in question concluded that their use in all food categories was safe within the acceptable daily intake (ADI) (Herrman and Younes, 1999). In ten years since note 161 was introduced, no discussion was initiated on hazards associated with the use of sweeteners in foods other than those with significant energy reduction or no added sugars. If the risk was associated only with misleading consumers, no risk assessment was requested on potential hazards associated with such misleading.

Second, it was not discussed or argued why consumers could be misled in one country and not misled in the other when they purchase foods with sweeteners, leaving the most important question unanswered: *could foods with sweeteners be traded internationally if they are not foods with low calories or no added sugars? And if not, why would CAC adopt provisions that work against fair trade practices?*

For the sake of transparency, aligned with Codex principles, countries concerned with the use of sweeteners should have substantiated their arguments through the four-step risk assessment. In contrast, CAC should have communicated the risk estimates to all members, thereby allowing for a proper risk management decision.

Instead, ambiguous notes were preferred over proper risk analysis, which compromised trust that member countries may have in CAC documents.

In conclusion, CAC communication with external stakeholders suffers from inadequate technical and online resources that have been lacking the capacity to serve the large number of users. Delays and breakdowns in addressing CAC web addresses are frequent, while navigation in document collections and databases are cumbersome and misleading.

Besides, not all CAC documents are available in all UN languages, whose use is a mandatory condition for both FAO and WHO. While the Codex Secretariat undertakes considerable effort to improve this situation, 70 percent of the CAC guidelines and 85 percent of the codes of practice are not translated into all six languages. Committees are mostly run in English with verbal translation offered in French and Spanish at best. This is hardly acceptable for a global standard-setting body, which aims to provide regulations for worldwide use.

Expert advice and science-based decisions

Risk communication is not the only issue CAC must address in improving the transparency of its work. The most challenging task is to provide a solid scientific basis for adopting documents through the four-step risk assessment model, conducted by the joint FAO/WHO expert bodies and consultations.

The Procedural Manual specifically mentions what WHO/FAO bodies must address when requesting scientific advice. For the Codex Committee on Nutrition and Food for Dietary Purpose (CCNFSDU), the FAO/WHO Joint Expert Meeting on Nutrition (JEMNU) is named as the primary source of nutritional risk assessment advice for Codex Alimentarius.

In 2014, CCNFSDU engaged in the process of establishing a nutrient reference value (NRV) – a numerical value of daily intake for the purposes of nutrition labeling and relevant claims – for two polyunsaturated fatty acids: eicosapentaenoic acid (EPA) and docosahexaenoic acid (DHA). Both acids are found in fish and plant sources, and are attributed with various health benefits.

Following numerous discussions between 2014 and 2016, CCNFSDU and the associated working group formulated a proposal to establish the nutrient reference values-noncommunicable disease (NRV-NCD)[18] at 250 mg/day, in association with reducing the risk of coronary heart disease (CHD) mortality.

The proposal was based primarily on recommendations of the three previous WHO/FAO expert consultations. One of the experts stated (FAO and WHO, 2010):

> There is convincing evidence that fish consumption and EPA plus DHA intake lower the risk of coronary heart disease mortality… The maximum positive effect from EPA + DHA was estimated to occur at 250 mg/day.

However, as the committee was heading toward the final approval of the proposal, some member countries expressed concerns that the FAO/WHO expert consultations were outdated (they were from 2000, 2003, and 2010). These concerns were broadly supported by the WHO, whose representatives pointed out that the most recent evidence showed no benefits of both acids to human health. They proposed to take into account the work of the WHO's Nutrition Guidance Expert Advisory Group (NUGAG), which, at the time, was working on systematic reviews of randomized clinical trials (RCTs) and prospective cohort studies that assessed the effect of unsaturated fatty acids, including EPA and DHA, on cardiovascular health.

---

[18] The NRV-NCD is a value associated with the reduction in the risk of diet-related noncommunicable diseases.

Most unexpectedly, despite opposition from several member countries, the WHO and CCNFSDU chair refused to request advice from JEMNU, as prescribed by the Procedural Manual, leaving the risk assessment to a policy-driven WHO expert group.

After a year-long research, NUGAG reported back to the committee that the systematic review of RCTs found weak associations of EPA/DHA intake with CHD mortality, which, "were small and seemed to be driven by studies at moderate/high risk of bias. They were also compromised by the volume of missing data."

At the 2018 CCNFSDU meeting, the WHO insisted that the quality of the studies reviewed were low, and the committee was in no position to establish the NRV-NCD. Shortly after, the committee agreed to discontinue work on this topic, avoiding discussion that, in a later publication of their systematic review, the NUGAG authors stated that increasing EPA/DHA intake may slightly reduce CHD mortality (Abdelhamid et al., 2020).

NUGAG also told the committee that additional evidence from the systematic review and meta-analysis of nonclinical observational studies were of low quality. The systematic review found that higher EPA/DHA intakes were associated with a 9 percent reduced risk of fatal CHD events (the review covered studies with 349,586 participants).

In effect, the attempt to establish the NRV-NCD failed for two reasons. First, the WHO insisted on using the Grading of Recommendations Assessment, Development, and Evaluation (GRADE) that the organization uses to assess the quality of a body of evidence in developing WHO guidelines (WHO, 2014). GRADE discriminates against nonclinical trials, which have been widely used in nutrition science, insisting that only RCTs, used in medical science, can produce high-quality evidence. Evidence from non-RCTs or nonclinical observational studies are always rated as low quality in GRADE.

This approach was criticized by the WHO experts themselves, as a recently published study revealed (Gopinathan and Hoffman, 2018). In a set of thirty-five interviews, the authors analyzed how the WHO guideline development process could be improved in the opinions of WHO experts. Through comments on the use of GRADE, the interviewees argued that consistent results from multiple well-designed observational studies could lead to a more favorable rating of the quality of the evidence, especially when interventions and policies could not be tested with RCTs due to ethical, legal, or logistical reasons. Thus, rating the quality of evidence collected in observational studies as low in environments where clinical trials are inapplicable, leads to biased conclusions and recommendations.

NUGAG experts may not have been aware that the complexity of running pharma-designed RCTs for foods requires a prolonged timeframe necessary for nutrients to show their effect in the human body. Unless participants are placed in metabolic wards, enforcing a high degree of intake compliance, RCTs for foods are conducted in free-living conditions with participants left without control over their dietary choices. Therefore, low compliance rates of most food-related RCTs consistently reduce the quality rating of study evidence (Jones, 2013).

WHO interviewees emphasized the need for guidelines about health systems and public health interventions to consider evidence generated by nonrandomized study designs and observational studies, especially in order to produce guidance about how to implement and scale-up interventions.

The second reason was concerned with the interpretation of data. The NUGAG systematic review concluded that, "167 people will need to take PUFA supplements for around four years each so that one of those people avoids a coronary heart disease event."

Authors argued that, compared to statin therapy that was associated with 4 percent mortality reduction (1 in 25 people taking statins for five years may not die from a cardiovascular event), it theoretically should take 1,000 people to take foods containing EPA/DHA for five years each for one person to avoid dying from CHD.

The view that the health benefits from food should be compared to the efficacy of the pharmaceutical drugs demonstrates that health and medical professionals of most respected scientific institutions drive global food regulation in the wrong direction. After all, in the United States alone, in 2017, CHD killed 365,914 people (Benjamin et al., 2019), which translates to one CHD death per 1,000 people. Even if EPA/DHA daily intake is said to take five years to save a single life among 1,000 Americans, this intake still reduces mortality rate from CHD by 20 percent. This is rather disappointing that NUGAG experts failed to notice these statistics.

The EPA/DHA example shows how poorly designed risk assessments delegated without observing requirements of the Procedural Manual yielded risk management decisions that contradicted earlier WHO/FAO recommendations and many national dietary guidelines around the world that advise consumers to increase EPA/DHA intake to reduce cardiovascular risks.

The CCNFSDU's decision also undermined CAC efforts to ensure transparency in Codex decision-making processes and highlighted the Commission's dependence on parent organizations, which are driven by political and economic agendas for their stakeholders.

Consensus and fair trade

The Procedural Manual reads that no decision should be brought to the Commission for approval by subsidiary bodies until consensus was achieved at the technical level.

There is a separate section in the Procedural Manual that lists mea-sures to be undertaken by chairs to facilitate consensus in the adoption of standards. While these recommendations are helpful, in effect, CAC is limited in its options to arrive to a full agreement. Failures to reach consensus are rather frequent, with countries expressing reservations[19] for decisions to approve or amend documents. It is also common that deci-sions agreed by subsidiary bodies are disputed and not approved at CAC annual meetings.

In 2012, the Codex Committee on Contaminants in Foods (CCCF), on request from several member countries, agreed to request an expo-sure assessment of cadmium (Cd) from cocoa and cocoa products by the JECFA.[20] Cd is a well-known toxic element with a high rate of soil-to-plant ratio (Satarug, Vesey, and Gobe, 2017). The major food categories that con-tribute to the Cd exposure are rice and grains, shellfish and seafood, meat including edible offal, and vegetables.

JECFA concluded that total Cd exposure, including for frequent con-sumers of cocoa and cocoa products, such as children, was not a safety concern. Still, cocoa-producing countries from Latin America, while agreeing that Cd was not a health concern, thought that the lack of an ML for Cd in cocoa and its products could threaten their cocoa exports (Codex Committee on Contaminants in Foods, 2014).

Cocoa grown in Latin America had considerably high levels of natu-ral Cd content compared to the same commodities produced in Africa and South East Asia. At the same time, many cocoa-importing countries established MLs for Cd, being aware of the high toxicity of the metal.

In 2014, the EU adopted new levels for Cd in cocoa and chocolate (The European Commission, 2014) based on its own safety assess-ment produced by the European Food Safety Authority. With 38 per-cent of cocoa beans grown worldwide ground in Europe and Russia (International Cocoa Organization, 2014), the EU levels were highly rele-vant to the CCCF discussion. As Table 2.2 shows, for chocolates contain-ing less than 50 percent of cocoa solids, the most consumed chocolate varieties in the confectionery market, naturally occurring levels of Cd in chocolates produced from Latin American cocoa were higher than MLs adopted in the EU. At the same time, African and Asian levels were orders of magnitude lower. Naturally, Latin American countries voiced their concern that the EU levels simply deny Latin American cocoa from entering the EU market.

---

[19] Reservations are not defined in the Procedural Manual. It is understood that countries express their reservations to state that they will not be bound to the decision taken by CAC.

[20] JECFA represents the world's top food safety experts who exercise the most stringent views on food safety ingredients and contaminants.

Table 2.2 Comparison of cadmium levels naturally occurring in chocolates produced from Latin American cocoa and levels set in the EU and proposed in CAC, mg/kg

Chocolate	Cd in LatAm cocoa	ML proposed by CAC	2014 EU levels
Milk chocolate with <30 percent total dry cocoa solids	0.25	0.3	0.2
Chocolate with <50 percent total dry cocoa solids; milk chocolate with ≥30 percent total dry cocoa solids	0.35	0.5	0.3
Chocolate with ≥50 percent total dry cocoa solids	0.61	0.8 (adopted)	0.8

In 2018, CAC approved ML of 0.8 mg/kg for chocolate containing 50–70 percent of cocoa, which was exactly the EU level. The value was also acceptable for Latin American countries.

Reassured by this success, in 2019, Latin American countries pushed hard to advance the ML of 0.3 mg/kg for chocolate with cocoa solids below 30 percent and 0.5 mg/kg for chocolates with cocoa solids between 30 and 50 percent. Both were higher than the EU levels, and the proposal was met with strong resistance from the EU.

At the emotional 2019 CAC session in Geneva, the chair insisted that CAC had to approve the ML of 0.3 mg/kg for chocolates with cocoa contents below 30 percent, as CCCF already approved it. The value was ultimately approved, but the EU still objected, insisting that children mostly consumed low-cocoa chocolates and Cd levels above the EU level of 0.2 mg/kg posed a health threat to these most vulnerable population groups.

It remains to be seen how the Codex value for Cd in chocolate approved by CAC without agreement of the EU, major cocoa and chocolate consumer in the world, will be put into practice. At the same time, it must be noted that CCCF discussions have been ongoing for six years, ignoring continuous assurances from JECFA that their assessment clearly showed that Cd contamination was of no concern both in cocoa and chocolate for all population groups including children.

The Cd in CCCF example demonstrated how the economic interests of member countries pushed CAC to an unknown territory of trade disputes where science-based risk analysis was powerless and irrelevant. JECFA expert authority was completely undermined by countries that sought to establish MLs when there was no safety concern, and by countries that insisted that their internal regulations have to be unconditionally observed by all parties, including CAC. Neither the JECFA nor the CCCF could have done better to reduce the transparency of Codex Alimentarius and the global food system in general.

Conclusion

It has been warranted by FAO and WHO that foods conforming to Codex Alimentarius standards should be freely bought and sold in the international market without compromising the health or interests of consumers. In this context, CAC contributes to the transparency of the global food system by providing science-based and consensus-based standards recognized by all participants of the food supply network. Since CAC's first meeting in 1963, member countries and observers have been engaging in CAC meetings with full trust that the Codex Alimentarius represents the best possible global effort to protect consumers and fair trade.

CAC focuses primarily on international trade; thus, it must ensure that foods remain safe globally irrespective of their origin or destination. The organization also contributes to the risk reduction of food-borne diseases and prevention of global food-related health issues, e.g., the AMR and NCDs.

CAC's Procedural Manual builds an extensive legal framework that encompasses all aspects of the food safety standard-setting process, focusing on balancing interests of all parties involved. Nevertheless, as real-life examples in this chapter demonstrated, CAC members, parent organizations, and external stakeholders often use CAC as a platform to pursue their own short-lived priorities, which compromise CAC fundamental objectives and damage its global reputation.

In summary:

- CAC2 consensus-based and science-based risk analysis approach is designed to produce a best-of-class food safety regulation for the whole world.
- Protectionism remains the key issue in international food trade and CAC clearly lacks efficient tools to advance fair trade practices.
- Certain flaws in CAC structure and procedures allow stakeholders to drive meeting agendas and discussions away from global priorities.
- Low technical capabilities limit CAC ability to reach broad worldwide audiences and establish effective risk communication with stakeholders, including consumers.

In the current environment, CAC seems to be lacking an independent supervisory function that would monitor the integrity of the decision-making process and ensure more responsive governance of the CAC executive and subsidiary bodies. Such body, reminiscent to a corporate board of directors, should be tasked with forcing compliance with the Procedural Manual while remaining independent from parent organizations, FAO and WHO, Codex Secretariat, and chairpersons.

References

Abdelhamid, Asmaa S., Tracey J. Brown, Julii S. Brainard, Priti Biswas, Gabrielle C. Thorpe, Helen J. Moore, Katherine H.O. Deane, et al. 2020. "Omega-3 Fatty Acids for the Primary and Secondary Prevention of Cardiovascular Disease." *Cochrane Database of Systematic Reviews* February. Wiley. Doi:10.1002/14651858.cd003177.pub5.

Benjamin, Emelia J., Paul Muntner, Alvaro Alonso, Marcio S. Bittencourt, Clifton W. Callaway, April P. Carson, Alanna M. Chamberlain, et al. 2019. "Heart Disease and Stroke Statistics—2019 Update: A Report from the American Heart Association." *Circulation* 139 (10). Ovid Technologies (Wolters Kluwer Health). Doi:10.1161/cir.0000000000000659.

Beulens, Adrie J.M., Douwe-Frits Broens, Peter Folstar, and Gert Jan Hofstede. 2005. "Food Safety and Transparency in Food Chains and Networks Relationships and Challenges." *Food Control* 16 (6). Elsevier BV: 481–486. Doi:10.1016/j.foodcont.2003.10.010.

Cocoa Market Outlook Conference 2015. London: The International Cocoa Organization. https://www.icco.org/wp-content/uploads/1-Laurent-Pipitone-Cocoa-Market-Outlook-2015.pdf.

Codex Committee on Contaminants in Foods. 2014. "Proposed Draft Maximum Levels for Cadmium in Chocolate and Cocoa-Derived Products: CX/Cf 15/9/6." New Delhi, India.

Codex Committee on Food Additives (CCFA). 2009. Report of the Forty-First Session of the Codex Committee on Food Additives: ALINORM 09/32/12.

FAO and WHO. 2010. *Report of Joint FAO/WHO Expert Consultation on the Risks and Benefits of Fish Consumption: FAO Fisheries and Aquaculture Report No. 978.* Rome. http://www.fao.org/3/ba0136e/ba0136e00.pdf.

FAO. 2020. "FAOSTAT." http://www.fao.org/faostat/en/#data/QV.

FAO. 1961. *Report of the Conference of FAO: Eleventh Session, 4–24 November 1961.* Rome. http://www.fao.org/3/x5572E/x5572e00.htm#Contents.

FAO. 2018. "Understanding Codex." Rome. http://www.fao.org/3/CA1176EN/ca1176en.pdf

FAO and WHO. 2011. *FAO/WHO Guide for Application of Risk Analysis Principles and Procedures during Food Safety Emergencies.* Rome: FAO/WHO. http://www.fao.org/3/ba0092e/ba0092e00.pdf.

Gardner, T.A., M. Benzie, J. Börner, E. Dawkins, S. Fick, R. Garrett, J. Godar, et al. 2019. "Transparency and Sustainability in Global Commodity Supply Chains." *World Development* 121 (September). Elsevier BV: 163–177. Doi:10.1016/j.worlddev.2018.05.025.

Gopinathan, Unni, and Steven J. Hoffman. 2018. "Institutionalising an Evidence-Informed Approach to Guideline Development: Progress and Challenges at the World Health Organization." *BMJ Global Health* 3 (5). BMJ: e000716. Doi:10.1136/bmjgh-2018-000716.

Hamer, John. 2013. *The Falsification of History: Our Distorted Reality.* Rossendale: Rossendale Books.

Herrman, J.L., and M. Younes. 1999. "Background to the Adi/Tdi/Ptwi." *Regulatory Toxicology and Pharmacology* 30 (2). Elsevier BV: S109–S113. Doi:10.1006/rtph.1999.1335.

InterAcademy Partnership. 2018. *Opportunities for Future Research and Innovation on Food and Nutrition Security and Agriculture. The Interacademy Partnership's Global Perspective: Synthesis by IAP Based on the Four Regional Academy Network Studies.* www.interacademies.org.

Joint FAO/WHO Codex Alimentarius Commission. 1963. *Report of the First Session: Ref. N° Alinorm 63/12 July 1963.* Rome. http://www.fao.org/input/download/report/3/al63_12e.pdf.

Joint FAO/WHO Codex Alimentarius Commission. 2018. *Procedural Manual: 26ᵗʰ Edition.* Rome: Food and Agriculture Organization of the United Nations.

Joint FAO/WHO Codex Alimentarius Commission. 2019a. *General Standard for Food Additives: CODEX Stan 192-1995.* Rome: Food and Agriculture Organization of the United Nations.

Joint FAO/WHO Codex Alimentarius Commission. 2019b. "Codex Strategic Plan 2020–2025." http://www.fao.org/3/ca5645en/CA5645EN.pdf.

Jones, Peter. 2013. *Best Practices for Food-Based Clinical Trials: Guidance for Planning, Conducting and Reporting on Human Studies to Support Health Claims.* Ottawa, Ontario: Minister of Agriculture and Agri-Food Canada.

Kuenzel, David J. 2020. "WTO Tariff Commitments and Temporary Protection: Complements or Substitutes?" *European Economic Review* 121 (January). Elsevier BV: 103344. doi:10.1016/j.euroecorev.2019.103344.

Melo, Oscar, Alejandra Engler, Laura Nahuehual, Gabriela Cofre, and José Barrena. 2014. "Do Sanitary, Phytosanitary, and Quality-Related Standards Affect International Trade? Evidence from Chilean Fruit Exports." *World Development* 54 (February). Elsevier BV: 350–359. doi:10.1016/j.worlddev.2013.10.005.

Peraica, M., B. Radić, A. Lucić, and M. Pavlović. 1999. "Toxic Effects of Mycotoxins in Humans." *Bulletin of the World Health Organization.* 77 (9). World Health Organization: 754–766. https://pubmed.ncbi.nlm.nih.gov/10534900 https://www.ncbi.nlm.nih.gov/pmc/articles/PMC2557730/.

Pipitone, Laurent. 2015. *Cocoa supply & demand: what to expect in the coming years?*

Robertson, Jeffrey, and Warwick Funnell. 2012. "The Dutch East-India Company and Accounting for Social Capital at the Dawn of Modern Capitalism 1602–1623." *Accounting, Organizations and Society* 37 (5). Elsevier BV: 342–360. doi:10.1016/j.aos.2012.03.002.

Satarug, Soisungwan, David A. Vesey, and Glenda C. Gobe. 2017. "Current Health Risk Assessment Practice for Dietary Cadmium: Data from Different Countries." *Food and Chemical Toxicology: An International Journal Published for the British Industrial Biological Research Association* 106 (Pt A): 430–445. doi:S0278-6915(17)30321-6 [pii].

Smith, D.F. 2007. "Food Panics in History: Corned Beef, Typhoid and 'Risk Society'." *Journal of Epidemiology & Community Health* 61 (7). BMJ: 566–570. doi:10.1136/jech.2006.046417.

The European Commission. 2014. "COMMISSION Regulation (Eu) No 488/2014 of 12 May 2014 Amending Regulation (Ec) No 1881/2006 as Regards Maximum Levels of Cadmium in Foodstuffs." https://eur-lex.europa.eu/legal-content/EN/TXT/PDF/?uri=CELEX:32014R0488&from=en.

The United Nations. 2015. "Transforming Our World: The 2030 Agenda for Sustainable Development." https://undocs.org/en/A/RES/70/1.

The United Nations. 2019. "Population Dynamics. Total Population by Sex." Department of Economic and Social Affairs. https://population.un.org/wpp/DataQuery/.

Vojir, Franz, Erwin Schübl, and Ibrahim Elmadfa. 2012. "The Origins of a Global Standard for Food Quality and Safety: Codex Alimentarius Austriacus and Fao/Who Codex Alimentarius." *International Journal for Vitamin and Nutrition Research* 82 (3). Hogrefe Publishing Group: 223–227. doi:10.1024/0300-9831/a000115.

World Bank. 2020. "World Bank Open Data." Accessed February 18. https://data.worldbank.org/.

WHO. 2014. *WHO Handbook for Guideline Development.* Geneva: World Health Organization.

FAO and WHO. 1950. *Joint FAO/WHO Expert Committee on Nutrition: Report on the First Session, Geneva, 24–28 October 1949.* World Health Organization Technical Report Series; No. 16. World Health Organization.

World Trade Organization. 1995a. "Agreement on Technical Barriers to Trade: 1868 U.N.T.S. 120." Accessed 1 March 2020. https://www.wto.org/english/res_e/publications_e/tbttotrade_e.pdf.

World Trade Organization. 1995b. "The WTO Agreement on the Application of Sanitary and Phytosanitary Measures (SPS Agreement)." World Trade Organization. https://www.wto.org/english/tratop_e/sps_e/spsagr_e.htm#fnt1.

chapter three

Agroecology in food system transparency: Labels and certifications

Jennifer Williams Zwagerman

Contents

EDITORS' NOTE: FAO'S 2030 AGENDA FOR AGROECOLOGY

As Zwagerman aptly notes in Chapter 3, the Food and Agriculture Organization (FAO) has published a multitude of initiatives to support agroecology with the quintessential need to combine knowledge and advances to further the goals of food system transparency. Complementary to Chapter 2, Zwagerman explains how agroecology fits into food system transparency.

Just as Petrenko and Tutelyan's chapter contextualizes food system transparency with trade and free markets, Jennifer Williams Zwagerman contextualizes food labeling transparency, here, with agroecology. Loosely, agroecology fits into yet another set of

Sustainable Development Goals (SDGs) by combing agriculture and ecology with the goal to achieve resilient food production.

As a brief note for our readers and by suggestion of the author of Chapter 3, the FAO's efforts to support agroecology are summarized in the mission statement for the "Scaling Up Agroecology Initiative: Transforming Food and Agricultural Systems in Support of the SDGs."[1] The goals of the mission are in a "transformative spirit of the 2030 Agenda," where the approach shall "harnes[s] a range of sustainable practices and policies, knowledge and alliances to achieve equitable and sustainable food systems in support of the SDGs."[2]

Specifically, the FAO outlines how agroecology embraces the spirit of the 2030 Agenda:

- *Agroecology helps to achieve multiple objectives through integrated practices, supported by coherent cross-sectoral policies.* Agroecology addresses the environmental, economic, and social dimensions of agri-food systems. It seeks innovative and holistic solutions to the complex and interrelated challenges of poverty, hunger and malnutrition, rural abandonment, environmental degradation, and climate change.
- *Agroecology places people at the center.* Agroecology empowers people to be the critical agents of change in the transformation of their food systems. It recognizes and brings together knowledge and experiences of diverse actors – including, women, youth, food producers, traders, consumers, policymakers, scientists, and citizens.
- *Agroecology contributes directly to multiple SDGs.* The eradication of poverty (1) and hunger (2), ensuring quality education (4), achieving gender equality (5), increasing water-use efficiency (6), promoting decent jobs (8), ensuring sustainable consumption and production (12), building climate resilience (13), securing sustainable use of marine resources (14), and halting the loss of biodiversity (15).[3]

In Chapter 3, Zwagerman further details how food labeling programs, and the information they convey or fail to convey, play a critical role in achieving these goals.

[1] http://www.fao.org/3/I9049EN/i9049en.pdf
[2] http://www.fao.org/3/I9049EN/i9049en.pdf
[3] http://www.fao.org/3/I9049EN/i9049en.pdf

Introduction

Consumers are increasingly demanding more information about the food they eat, particularly that purchased for in-home consumption. Food manufacturers are responding to this demand by adding more and more information to the food label in the name of transparency. This information goes beyond the required nutrition information and traditional health claims, instead providing voluntary information about the origin of food products, from how it was raised, food facility standards, and a host of other types of information. The food industry claims that this increases transparency and provides consumers more information about where the food comes from, how it was produced, and information regarding the environmental impacts of various food products.

As consumer demand and interest has evolved and expanded from a focus on the personal impact of food choices (i.e., health and nutrition) to a broader interest in the environmental and sustainability issues related to food production, the types and varieties of claims in this area increased dramatically. While consumers indicate they rely upon this information that falls under the umbrella of agroecology, often used interchangeably with "sustainability" by consumers and food manufacturers, these claims are largely unregulated, unverified, and undefined. While principles of agroecology have entered our mainstream lexicon, particularly over the past decade, numerous challenges exist that undermine a consumer's ability to trust, verify, or ensure that claims made on food products are accurate or effective in addressing the underlying concerns. As we look at some of these challenges, including a lack of universal definition, we will also explore opportunities and difficulties within current labeling and regulatory systems to develop a system that provides consumers clear and accurate information related to the agroecological impact of food choices.

Agroecology: Beyond sustainability

Before we discuss the evaluation of agroecological claims on the food label, we first need to address the challenge of defining agroecology. Agroecology itself is a term that evolved over the years, since scientists and academics first started to look at ecology and agronomy as sciences that could be merged and studied together.[4] Once viewed simply as the ecology of agriculture,[5] the word now refers not to just ecological principles of agriculture, but includes social, political, economic, cultural,

---

[4] Michel P Pimbert, Global Status of Agroecology: A Perspective on Current Practices, Potential and Challenges, Economic & Political Weekly No. 41 (Oct. 13, 2018) p. 53
[5] *Id.*

ethical, and technological/scientific considerations.[6] Agroecology has gained international recognition, driven in part by increased understanding of the environmental impacts of agriculture, particularly industrial agriculture, and enhanced concern over inequalities within the food system.

Agroecology definitions

In addition to an evolving definition, the term agroecology has taken on different meanings in different regions of the world. The broader and more diverse the definition of agroecology, the more difficult it is to determine if food products are accurately and transparently making agroecological claims on the label.

The lack of a clear, concise definition of both the terms agroecology and sustainability has not stopped food companies from responding to consumer demand and adding various labels making these claims to products. The Ecolabel index was tracking over 450 ecolabels in approximately 200 countries and in 25 industries, as of mid-2020.[7] Eco-label is defined as "A sign or logo that is intended to indicate an environmentally preferable product, service or company, based on defined standards or criteria."[8]

Author and professor, Gabriela Steier, defined several key terms relevant to determining how we evaluate and consider food label claims related to agroecology: **Agroecology** is the ecology of food systems.... **Sustainability** shall be defined as pertaining to a food system that maintains its own viability by using agroecological techniques that allow for continued reuse and holistic service to all components of food integrity.[9]

With few exceptions, the majority of agroecological labels currently in use focuses on the environmental impacts, short and long term, of a particular food item. Sustainability is the buzz word that grabs people's attention and resonates with consumers. For the purposes of this main discussion, we will focus on the principle that the basic tenant of "agroecology is the idea that agroecosystems should mimic the biodiversity levels and functioning of natural ecosystems...[s]uch agricultural mimics … can be productive, pest-resistant, nutrient-conserving, and resilient… [with] no 'waste'; nutrients are recycled indefinitely."[10] While the majority

[6] *Id.*

[7] http://www.ecolabelindex.com/

[8] http://www.ecolabelindex.com/glossary/

[9] Steier, Gabriela, Advancing Food Integrity: GMO Regulation, Agroecology, and Urban Agriculture, *Glossary* p. 223–224 (2018, CRC Press)

[10] Michel P Pimbert, Global Status of Agroecology: A Perspective on Current Practices, Potential and Challenges, Economic & Political Weekly No. 41 (Oct. 13, 2018) p. 52

of examples focuses on environmental-related claims, we will reference others that touch on some of the more cultural, ethical, economic, and social aspects of agroecology.

Agroecology and the food label

Regulation basics: Would I lie to you?

The food label has long been a source of mystery for consumers, despite the fact that many of its elements are highly regulated by the Food and Drug Administration (FDA) in the United States, and similar agencies across the globe. The majority of regulated claims on the label, particularly in the United States, is related to health and nutrition.[11] Agroecological claims on the label typically fall outside of FDA's strict regulations and definitions. While many of the terms and claims that will be discussed here are not defined by FDA (or its counterpart for meat and poultry product regulations, the United States Department of Agriculture (USDA)), that does not mean that claims are wholly unregulated.

All elements of a label, words, symbols, numbers, etc., fall under the prohibition against mislabeling that is contained in the Federal Food Drug and Cosmetic Act (FFDCA). Specifically, the FFDCA prohibits "[t] he introduction or delivery for introduction into interstate commerce of any food … that is … misbranded" (21 U.S.C. § 331(a)). A product is deemed misbranded if the label is "false or misleading in any particular" (21 U.S.C. § 343(a)). That means that any agroecological (or other) claim made on a food label must be true, and it must not mislead consumers into thinking something that is inaccurate. The "false or misleading" aspect is extremely relevant to the idea of transparency. If a food item contains a claim that is true and not misleading, then a company should be able and willing to provide information to consumers to establish the veracity of the claim and to define what it is based upon. With many of the terms used in labeling claims undefined, that leaves much room for interpretation, and in turn, litigation.

For example, the term "natural" is often used to market food products, including some products using the term to refer to the product occurring in or from "natural ecosystems" like those in the definition of agroecology discussed above. However, FDA has declined to define the term natural, and litigation over the use of that term and whether its use is false or misleading relevant to the product in question has skyrocketed

---

[11] FDA, *Label Claims for Conventional Foods and Dietary Supplements, available at:* https://www.fda.gov/food/food-labeling-nutrition/label-claims-conventional-foods-and-dietary-supplements

in the past two decades. One count found less than 20 such federal class action cases filed in 2008, rising to over 100 in 2011.[12] That does not include the numbers of state class action lawsuits brought under state consumer protection laws, with plaintiffs arguing that a product's claim of being "natural" was false or misleading due to its ingredients, processing, or other related issues.

Similarly, FDA has not defined the word "sustainable" when used on food labels, leading to similar litigation over its use on food products. Not only did FDA decline to define the term, but the Federal Trade Commission (FTC), responsible for regulation of food marketing and advertising, also refused to define the term in its guidance for the food industry.[13] When it comes to marketing, the FTC regulation is very similar to that of FDA's misbranding definition. Instead, the FTC focuses on consumer deception and the prevention of deceptive actions in commerce. In marketing and advertising, the FTC states that "a representation, omission, or practice is deceptive if it is likely to mislead consumers acting reasonably under the circumstances and is material to consumers' decisions."[14] The FTC developed Green Guides to provide assistance to those wishing to make environmental claims regarding products across a host of industries. However, when it came to the term "sustainable," the lack of a single consumer meaning led to discomfort in attempting to set a particular definition or standard for the term.

So, with these standards in place, can a consumer safely assume that any claim made on a label is a fair and accurate statement? The answer, in most cases, is sadly no, particularly when the claims being made are not verified or certified by a third party. While there is some risk in a company making a claim as to the food products environmental-friendliness that is perhaps less than clear or perhaps not as impactful as a consumer might think, it can be very hard to prove consumer confusion or deception when the terms itself can naturally lead to different interpretations.[15] While companies should have evidence to substantiate a claim prior to making it, that information does not need to be made public in most cases unless the claim is challenged through litigation, or becomes part of the public record through FTC or FDA actions (such as

[12] https://www.brookings.edu/wp-content/uploads/2016/06/Negowetti_Food-Labeling-Litigation.pdf
[13] https://www.crowell.com/NewsEvents/Publications/Articles/Natural-Sustainable-Risks-of-Using-Undefined-Terms-in-Food-Marketing
[14] https://www.ftc.gov/sites/default/files/attachments/press-releases/ftc-issues-revised-green-guides/greenguides.pdf
[15] https://www.crowell.com/NewsEvents/Publications/Articles/Natural-Sustainable-Risks-of-Using-Undefined-Terms-in-Food-Marketing

FDA warning letters and responses, which are public record and posted on the FDA website).

On point or missing the mark: Label and certification examples

In reviewing examples of various programs, certifications, and label claims, it is clear that transparency is not a defining principle of many of these programs. This is in part due to the complex nature of our food system, frequent supply chain changes, concern about trade secrets or company operations, and a lack of a true regulatory or certification system or agroecological claims or programs. Claims made by an individual company are often more difficult to verify as there is not a clear process in place for a consumer to visit the farm of an ingredient source, tour a facility, see prices paid for goods to ensure fairness, or in short, have access to a company's records, facilities, and suppliers.

Non-GMO project

The Non-GMO Project is a targeted third-party certification system focused on providing information to consumers to identify products that are derived from non-GMO ingredients. The program mission statement does touch on several tenets of agroecology, including ensuring the integrity of plant genetics, supporting traditional seed breeding, supporting "environmental health and ecological harmony, and protecting supply chains.[16] Formed in 2007 by two grocery stores, the nonprofit launched the first products into the marketplace with its verification symbols in 2010. Ten years later, more than 50,000 products and 3,000 brands have gone through the verification system. Its butterfly logo is well-known and is seen in stores internationally.

Consumers recognize this label, and a Consumer Reports study found that it was "the only 'highly meaningful' label for consumers looking to avoid GMOs."[17] In terms of transparency, there are many positive aspects. Products participating in this certification go through a third-party verification system with independent reviewers. The independent reviewers review information provided by the company related to the product, the ingredients, and the facility in which it is manufactured. Testing may be required in some instances to ensure that products meet the non-GMO standard. The standards themselves are public, posted on the nonprofit's

[16] https://www.nongmoproject.org/about/mission/
[17] https://www.consumerreports.org/cro/2014/10/where-gmos-hide-in-your-food/index.htm

website and reviewed regularly (version 15 was released in 2019). The standards publication is over forty pages long and includes extensive details on the program requirements for a wide variety of products and ingredients.

The non-GMO program is an example of a program that transparently provides its standards and definitions and processes to consumers and participants alike, meaning that all information is available for consumers to understand exactly what that certification means on the food label. This does require an assumption that the independence of the program administrators and reviewers exists, all protocols are followed, and audits are conducted as scheduled. These assumptions are necessary because while there is information provided to consumers about the standards and processes, individual product information is not provided.

While clearly an environmentally focused certification program, perhaps one that can be defined as a sustainable agricultural certification system under certain definitions, it is a stretch to view a program such as this as one that truly meets even the more limited definition of agroecology adopted for the purposes of this chapter. While this program/certification itself might not be considered agroecological, in many cases, the non-GMO Project logo will not be the only certification nor claim on the product. For consumers looking to purchase products that are not only non-GMO but also fair trade certified or raised in a humane fashion, they will have to look at other aspects of the label to determine if the products meet those additional standards.

USDA bioengineered foods label

The Bioengineered Foods labeling requirement is almost an offshoot or outgrowth of consumers who supported the non-GMO project and advocated for more information on the food label. The consumer "right-to-know" movement heavily focused on foods that included ingredients derived from genetic engineering, which under USDA definitions, is a very specific form of gene modification.[18] When genetically engineered (GE) crops first entered the food supply in the 1990s, FDA's stance was that in general there was no need to label foods contained GE ingredients as the foods were substantially similar to the non-GE counterparts.[19] Labeling was only required if there was a change in the GE product that changed

---

[18] USDA Agricultural Biotechnology Glossary, available at: https://www.usda.gov/topics/biotechnology/biotechnology-glossary; FDA Guidance for Industry, *Voluntary Labeling Indicating Whether Foods Have or Have Not Been Derived from Genetically Engineered Plants* (March 2019)

[19] FDA, *Food for human consumption and animal drugs, feeds, and related products: Foods derived from new plant varieties*, 57 Fed. Reg. 22984 (1992)

its nutritional profile or raised some sort of unexpected health concern.[20] Many consumers were unhappy and the Right-to-Know Movement took off, with state initiatives being a primary focus. While legislation was successful in a few states, the food industry rallied against this movement, defeating several state ballot initiatives and spending hundreds of millions of dollars in the process. Vermont's passage of required GE labeling[21] though was a turning point for both the Right-to-Know Movement and the food industry. While litigation was immediately filed challenging Vermont's labeling requirement, many food companies began voluntarily adding information to food labels indicating to consumers that some (or all) ingredients were derived from GE ingredients. At the same time, the industry began to lobby for federal legislation to address this issue, rather than a series of state laws with different regulations and requirements. The result was the National Bioengineered Food Disclosure Law, requiring disclosure of bioengineered ingredients and preempting state laws such as Vermont's.[22]

This is a federal labeling requirement, but while FDA is responsible for regulating the majority of food labels in the United States, USDA was tasked with implementing this program. The disclosure standards were released in late 2018 and formal implementation began in 2020. As a federal program, transparency is present in that much of the program itself is a matter of public record. The public and industry had the opportunity to comment on the proposed regulations before finalized. Yet consumers have many questions as to if the information being provided is truly representative of the goal of the law: provide information to consumers about the presence of bioengineered ingredients in food products. The federal standard limits disclosure to detectable materials modified through particular scientific techniques. It does not include animal products derived from animals that consumed bioengineered feed. It does not apply to foodservice establishments. Information is provided, but not always in a neat, orderly, or understandable way for consumers to determine standards, verification, and applicable products.

There is a system in place for consumers to file a complaint if they believe a product that is required to disclose bioengineered ingredients is not doing so appropriately, but consumers do not have access to product information or much of the technical data submitted by companies. From a transparency standpoint, information on the labeling requirement standards is available and the assumptions we make in this case are that

---

[20] *Id.*
[21] 9 V.S.A. §§ 3041-3048 (2014), available at https://legislature.vermont.gov/Documents/2014/Docs/ACTS/ACT120/ACT120%20As%20Enacted.pdf.
[22] https://www.ams.usda.gov/sites/default/files/media/Final%20Bill%20S764%20GMO%20Discosure.pdf

the government will enforce the law and regulations. However, from an agroecology standpoint, questions remain as to the effectiveness of this type of requirement, with its many exceptions and tolerance threshold. Are consumers truly able to trust that a product that does NOT display the bioengineered ingredients symbol is free of all such ingredients? Or do they understand that not having the label does not always equate to no bioengineered products as allowed pursuant to the regulations? This type of requirement raises inherent questions about transparency and confusion among consumers.

Genetic engineering and bioengineered foods, more specifically that GMOs more broadly, are relevant to questions of agroecology. Many argue they contribute to the increasing monoculture, particularly in countries such as the United States with high percentages of industrialized agriculture.[23] Other agroecological concerns tied to GE include economic (access to technologies), cultural (loss of indigenous knowledge and practices), environmental (biodiversity, mutations, pollinators), and health (food security).[24]

B Corps

B Corporations (B Corps) are one of the most comprehensive certification programs in this review, in that it is a program open to companies from all industries. The first B Corps were certified by the nonprofit B Lab in 2007, with more than 2,500 B Corps in over 50 countries as of 2020.[25] To become a certified B Corps, standards must be met related to social and environmental performance, meaning that the company's impact on employees, environment, customers, and community as a whole are evaluated at every level of the supply chain. Well-known companies such as Patagonia, New Belgium Brewing, Ben & Jerry's, and Toms have become B Corps, and many are starting to promote this certification to consumers.

The certification process varies depending on a company's size and structure, but are comprehensive and thorough, and in addition to impact evaluation includes various legal requirements. Those wishing to become a certified B Corps complete an assessment and then proceed through the multistep verification process. Ten percent of those being recertified each year are randomly selected for a site review, as part of the nonprofit's goal of maintaining transparency and credibility.

Transparency is a key theme for this program. The Impact Reports for all B Corps are publicly available for review. Requirements for

---

[23] Food Labeling and the Environment, 34 J. Envtl. L. and Litig. 1
[24] Food Labeling & GMOs, International Law and Indigenous Peoples, 30 Pace Int'l L. Rev. 1
[25] https://bcorporation.net/about-b-corps

transparency may also be enhanced for certain types of companies or in certain situations. The publicly available reports include information about a company's impact area scores and a breakdown, but it is not completely transparent and clear and to how the scores are determined and what they are based upon. It is possible to see where a company is strong in terms of impact, and where there are lower scores and perhaps room to improve. Some company may voluntarily provide the entire assessment on their own, which does provide the clarity missing from the Impact Report. B Lab, the coordinating nonprofit, also offers an opportunity for public complaints related to intentional misrepresentation of claims in the application process and for violations of the core values that B Corps member agree to support.

The B Corps stated goals, when applied to the food system, align in many ways with those of agroecology. Companies certified under this program are indicating a willingness to work "toward reduced inequality, lower levels of poverty, a healthier environment, stronger communities, and the creation of more high quality jobs with dignity and purpose."[26] As of mid-2020, just shy of 60 companies in the "Agricultural Services" industry are B Corps, the majority located outside of the United States, and over 375 B Corps are in the food and beverage industry, with just under half in the United States. Expansion of B Corps within the agricultural sector is an exciting opportunity. The comprehensive goals and organization of this program, combined with its transparency in terms of working with members to set goals, strive to improve, and collaborate to make social change, is a necessary factor in creating change.

Fair trade certified

The Fair Trade Certification program focuses on social, environmental, and economic standards, with the goal of supporting sustainable development for communities.[27] Fair Trade Certification standards exist in four areas: Agricultural Production, Factory Standard (Apparel & Home Goods), Capture Fisheries, and Trade. All of the standards and other materials are available in multiple languages, making them accessible to those most directly impacted by the Certification's stated benefits.

The certification process itself is done through a nonprofit utilizing independent auditors to work with participants and ensure standards are met. The stated focus: "social and environmental responsibility, meaning the well-being of the humans growing and processing products and the

---

[26] https://bcorporation.net/about-b-corps
[27] https://www.fairtradecertified.org

environment around them."[28] Requirements for certification are grouped into six key areas: "empowerment; fundamental rights at work; wages, working conditions and access to services; biodiversity, ecosystem function, and sustainable production; traceability and transparency; and internal management system."[29]

For a consumer looking to determine what this label claim means, the information is readily available and easy to read and understand. Its comprehensive, but the supplemental information provided ensures any potential confusion is minimized. Transparency is also indicated on the label itself. Use of the Fair Trade Certified seal is only allowed if the product is made up of 100 percent certified fair trade ingredients.[30] A modified version of the seal, identifying the fair trade ingredients individually or in general can be used if 20 percent or more of the product ingredients are certified. If less than 20 percent of the ingredients are certified fair trade, no seal can be used but a statement indicating particular ingredients are fair trade certified can be included on the label. While the percentages and differences in seals may not be readily definable without research, these limitations do prevent misleading use of the fair trade seal such that a product may be purchased under the assumption it is 100 percent fair trade ingredients, when in fact only 8 percent of the ingredients were fair trade certified.

The goals and actions of this program are highly in line with principles of agroecology. Transparency and validation of transactions are key aspects of the program and with trust in the third-party program, appears to be substantive. While the label claim is designed to drive interest and demand for these types of products, it is also a program that has immediate impact in that the standards needing to be met to receive the Fair Trade Certification provide real change and value to a vulnerable population. While many of the agricultural production certified industries involve commodities raised outside the United States, the trade standards have a direct impact on U.S. food companies and consumers, and through purchase of fair trade certified ingredients, on the agricultural producers as well.

USDA organic program

Like the B Corps Certification program, the USDA Organic program applies to products beyond food, including clothing and cosmetics. The National Organic Program (NOP) was established in 1990, with a primary

---

[28] https://www.fairtradecertified.org/business/standards/agricultural-production-standard
[29] https://www.fairtradecertified.org/business/standards/agricultural-production-standard
[30] https://www.fairtradecertified.org/business/standards/trade-standard

goal of providing assurance to consumers that consistent, uniform standards would exist and be applied to all products bearing the label, enhanc-
ing confidence in the organic label.[31]

For a food product to be labeled organic, only approved substances
can be used in the production process. These regulations state that
"organic agriculture [is] the application of a set of cultural, biological, and
mechanical practices that support the cycling of on-farm resources, promote ecological balance, and conserve biodiversity." Natural processes
and materials are requirements for organic production, from soil condi-
tioning to pest and weed management. Animal agricultural also has a
series of organic production practices concerned with living conditions,
access to pasture, health, feed (organic only), and animal origin.[32]

Farms must be certified by an accredited third-party agent, who
reviews all the materials and determines compliance. Annual reviews
and inspections are required as well.

Organic labels take different forms, depending on the percentage of
a product that is certified organic. The Certified Organic seal can only be
used if 95–100 percent of the ingredients are certified organic. If at least 70
percent of an item's ingredients are organic, the seal cannot be used, but a
statement can be made that the product is made with organic ingredients
and specify which ingredients are organic. All three of these types of label
statements require official certification and verification of production standards. However, a product with less than 70 percent organic ingredients can
still use the word "organic" on the label, without being certified, by simply
indicating in the ingredient list that the relevant ingredients are "organic."

The NOP has been the subject of much controversy over the years.
Allegations included that independent auditors were failing, consumers
were consumed over what organic meant in relation to health and the
environment, that approved and nonapproved substances were arbitrary,
and that organic production was actually worse for the environment.[33]
More recently, animal welfare regulations for organic livestock operations
were rescinded before they could go into effect, continuing the debate
over the scope of the NOP and if welfare standards were going beyond
the authority of the program.[34]

---

[31] NOP Preamble

[32] https://www.ams.usda.gov/sites/default/files/media/Organic%20Practices%20
Factsheet.pdf

[33] https://www.washingtonpost.com/business/2019/03/14/organic-food-industry-is-
booming-that-may-be-bad-consumers/; consumer report article on consumer confusion;
https://www.usnews.com/news/national-news/articles/2018-12-13/study-organic-food-
is-worse-for-the-climate-than-non-organic-food

[34] https://abcnews.go.com/US/trump-usda-withdraws-animal-welfare-regulation-
organic-farms/story?id=53745900

Transparency is not a key component of the NOP as a whole. USDA has many resources available that explain the program, the process, and the label, but the regulations themselves are difficult to understand should a consumer want to truly understand what is allowed or not-allowed as part of the NOP. Concerns over the veracity of the audit process and instances of false organic products also lessen consumer trust in the program and the information that is provided.[35]

Many questions are also raised as to the effect of the NOP on sustainability and the environment as organic production becomes more industrialized.[36] While regulations and certifications are transparent, what, if any, actual positive impact is occurring at a farm or regional level is not clear at all with the NOP, as well as many other programs tied to environmental and sustainability issues. From a broader agroecology standpoint, the NOP is one of the more limited examples discussed, as there is clearly a focus only on certain aspects of agricultural production and biodiversity. The rescission of the animal welfare standards and arguments that it was beyond the scope of the program highlight that limited scope.

Regenerative organic certification

The Regenerative Organic Certification (ROC) appears to have grown out of a dissatisfaction with the NOP as not being comprehensive or effective enough to drive true change in the industry and meet the needs socially or environmentally of a struggling industry. Instead of focusing primarily on environmental health, the ROC also incorporates animal welfare and social fairness as core tenet of the certification program, while also requiring NOP Certification. ROC conducted a pilot program that ended in early 2020, and is looking to officially launch the program more broadly in the near future. Many of the assessment and standards documents are not yet finalized, but the draft and archived versions indicate a tiered structure with three levels of certification. ROC appears to utilize existing certification programs, including NOP, Fair Trade Certified, and Certified Humane, allowing certification in approved programs to satisfy standards for soil health, animal welfare, and social fairness in the ROC program.[37] This unique approach helps address the concern than many certification programs focus on only very narrow aspects of agroecology, and can also conserve ROC resources by limiting the amount of independent verification that needs to occur. The inclusion of definitions in the standards guide also helps reduce confusion and uncertainty over the

---

[35] Cite two recent cases (Cedar Rapids/MO grain case and dairy case out of MO)
[36] waPo article
[37] https://regenorganic.org/wp-content/uploads/2020/06/ROC-Framework-June2020.pdf

use of terms that may have multiple meanings or vary in certain contexts (such as inhumane).

Despite its newness, I include it in the discussion as a contrast to the NOP. The framework of the ROC more closely falls under the umbrella of agroecology, as it sets standards that go beyond the soil and look to fairness and welfare as well. The ROC allies are also strong, including companies like Patagonia, Thrive Market, Rodale Institute, Horizon, and a host of other organizations. By supporting a program that goes beyond soil health, this recognizes that these additional components are necessary for agriculture to succeed and that transparency, through third-party verifications and audits, is required to provide assurance consumers and the supply chain.

What do we want to know: Transparency and agroecology labels

The current status of labeling when it comes to agroecological traits and ideals is stunted by the lack of a comprehensive system and cohesive definition to set standards. When it comes to consumers, many indicate interest in purchasing environment-friendly, sustainable, and fair products, but it is far from clear that the current programs have been successful in actually increasing demand for these types of products or improving the status of the food system as a whole and those that work within it.

Another constraint of many of the most common labeling and certification systems is the focus on one particular aspect of agroecology, without providing any context to the claim itself. The use of sustainability interchangeably with agroecology means that the focus is on the environment, and on one particular moment or product or process, and not the impact of our food system as a whole. The strongest programs are those that are more comprehensive and understand that success in one area, such as reducing the environmental footprint of a particular food item or company, can easily be outweighed if that occurs at the expense of those that produce the food.

The programs described above are just a few of the many, many examples of those that exist around the globe. The clutter on the food label poses a challenge to any new or existing labeling program, as label fatigue is common among consumers. The sheer amount of information presented to them is overwhelming, important material can easily be lost in the plethora of other symbols and statements, and claims that are strong and transparent can suffer when consumers begin to doubt claims due to a lack of transparency, misleading statements by a bad actor or a failure in the verification system.

When it comes to transparency, third-party certification systems offer the best path forward in that they provide a clear set of standards and an audit process. This audit process in particular, and the ability to clearly understand what is being certified, are central to the idea of ensuring transparency to consumers who review the label. Government certification programs also suffer from a general distrust of the government, limits on the ability to share information, and concerns with changing administrations and priorities that impact regulatory and enforcement budgets.

Beyond being transparent about what a claim or a symbol stands for on the package, there is a need to be transparent regarding the impact and success of these programs at both a short- and long-term level. Transparency requires information about goals, challenges, opportunities, and it requires collaboration between those working for the same goals. This type of transparency goes beyond what might be required by the FTC or FDA to ensure a claim is not misleading or deceptive. This is transparency that goes beyond the regulatory and legal realm into the social, moral, and ethical realm.

There are examples of programs that currently exist that appear to understand the importance of both aspects of transparency, as well as the need for a holistic look at the agroecological issues that need to be addressed for agriculture to succeed. Established programs such as B Corps that encourage growth through public disclosure of data and goal setting and new programs such as ROC, which builds upon already existing programs and pulls them into a more holistic certification program that focuses on driving change and evaluating results, showcase the potential for transparent information on the food label. Transparency does not require more words on the label, but clear and accessible information regarding the standards and verification process, while also evaluating the claims and the goals of the program to ensure progress.

chapter four

Agriculture, freedom of speech, and the birth of ag-gag

Autumn Johnson

Contents

EDITORS' NOTE: BURDENS ON DEMOCRACY: AG-GAG AND VEGGIE LIBEL LAWS

It is a slippery slope to penalize whistleblowers – and this is what ag-gag laws do as applied to farms. In its simplest form, these laws hamper free speech, as Autumn Johnson explains in her chapter. She aptly characterizes these laws as criminalizing the documentation of agricultural practices, such as filming in Concentrated Animal Feeding Operations (CAFOs). Importantly, the lack of information, or the obstruction of a free flow of information, hinders transparency in our food system. Just as reporters should have a seat at the table to inform the general public about political and cultural events, the same freedom of consumer information is necessary at farms to advance food system transparency.

    Another form of these free speech impediments is "veggie-libel" laws, also called food disparagement statutes. Gabriela Steier published an article about the conflicts of interests created by the hindrance of freedoms to criticize foods and this research continues to be pertinent.

When Oprah Winfrey exclaimed on national television that mad cow disease stops her "cold from eating another hamburger," she put veggie libel laws in the spotlight. CNN reported that the beef industry lost $12 million because of her 1996 show on dangerous foods, among which she counted beef. A lawsuit arose. In *Texas Beef Group v. Winfrey*, 201 F.3d 680, (5th Cir. 2000), the Fifth Circuit ruled that Howard F. Lymann, farmer and animal welfare activist, did not violate the Texas False Disparagement of Perishable Food Products Act when he stated on the Oprah Show that Bovine Spongiform Encephalopathy (BSE), a.k.a. mad cow disease, could "make AIDS look like the common cold." The "beef industrialist" sued Oprah Winfrey for allegedly defaming their products. At the heart of the case was the Texas False Disparagement of Perishable Food Products Act, an example of veggie libel laws. Little progress has been made since the show aired in 1996.

Veggie libel laws, also called food disparagement statutes, attempt to stop defamatory statements that could harm industrial food producers in various states: Louisiana, Idaho, Mississippi, South Dakota, Texas, Florida, Arizona, Oklahoma, North Dakota, Colorado, Alabama, and Ohio. These laws prohibit statements that may reveal the ugly truth about foods sold in supermarkets as being unhealthy or even revolting. Such statements are, of course, harmful to the industry. Large beef and pork producers lobbied those laws into being to prevent the critical but free discourse about their products at the cost of consumers' free speech. The idea that the large food corporations lobbied these laws into being in the very states that produce the staples of the fast food industry is suspicious. When free speech about fast food in those states is restricted, consumers are left questioning whether the companies are hiding something.

Friedrich Nietzsche once stated, "Not when truth is dirty, but when it is shallow, does the enlightened man dislike to wade into its waters." Many of the common perceptions about American staple foods are extraordinarily shallow. When Oprah Winfrey aired her series on dangerous foods, she made a considerable step in sparking consumers' interest. She encouraged the public to wade into the water and explore the origins of everyday foods. Fifteen years later, in February 2011, her campaign to take the "vegan challenge" and eat a purely plant-based diet, as advocated by Michael Pollan, best-selling author and professor of journalism at the University of California, Berkeley Graduate School of Journalism, offered an alternative to the largely unsustainable practices of the food-animal industry and unhealthy food choices: veganism. By choosing mostly unprocessed, sustainably grown, plant-based foods, consumers are empowered to change the industrialized food production around

— from a vice to a virtue.[1]

Not much progress has been made in lifting the shroud of secrecy that envelopes industrial food production. Even though foodies and the food movement have changed how information is shared via social media, there is not necessarily more information available due to legally questionable ag-gag laws. In light of current affairs, the 1964 landmark case of *New York Times v. Sullivan*, 376 US 254, 270 (1964), is more current now than ever before. Johnson does an exemplary job in contextualizing the evolution of ag-gag in the following chapter.

[1] Steier, Gabriela, From Veggie Libel Laws to Planetary Sustainability (2012). Juris Magazine, p. 18, Spring 2012. Available at SSRN: https://ssrn.com/abstract=2115547

Introduction: What is ag-gag?

Laws criminalizing the act of documenting activities related to agriculture, including video, photos, or scientific samples, have popped up in states all over the country. Ag-gag laws are often targeted at environmental or animal welfare advocates. These laws, commonly referred to as "ag-gag," are being challenged in the courts, often being overturned in whole or in part due to freedom of speech violations. While this area of the law is very much unsettled, this chapter attempts to explain the types of ag-gag laws, where they have been passed, and what the courts have said about them to date. The Animal Legal and Historical Center at Michigan State University offers the following summary:

Ag-gag laws generally fall into one of three categories or a combination of two or all types.

1. The first modern-day ag-gag laws are known as "agricultural interference" laws, and they ban recording images or sounds at industrialized farming operations without the owner's consent.
2. "Agricultural fraud" laws were also introduced early on; this kind of ag-gag law bans entering or applying for employment at industrialized farming operations under false pretenses.

3. "Rapid-reporting" legislation mandates that anyone who records an image or sound at an industrialized farming operation turn the recording over to authorities within a specified amount of time, usually twenty-four to forty-eight hours.[2]

While the third type, regarding the time to report, may seem innocuous, prosecutions of animal welfare violations typically require documentation over a substantial period of time. Having to report violations immediately prohibits collecting enough evidence to document a pattern of behavior. According to the Humane Society of the United States, "requiring that any violation be immediately reported [is] intended to prevent whistleblowers from establishing a pattern of abuse. Because of the power of the factory farming lobby, prosecutors will rarely file charges for farm animal cruelty unless a pattern has been established and the evidence is overwhelming."[3]

"Ag-gag" laws are also sometimes referred to as "farm protection laws" or "anti-whistle blower laws" (hereinafter, "ag-gag"). Other related statutes are sometimes called "eco-terrorism" or "animal enterprise interference" laws. Michigan State University keeps a table of all such laws by state.[4] Ag-gag laws vary from state to state.

The PEW Charitable Trusts, a nonprofit advocacy group, attributes the rise of ag-gag laws with animal welfare advocates successes in reforming the industry. "When animal welfare groups started about a decade ago to pay their employees to take jobs on farms to expose practices, the industry responded by pushing for what animal welfare advocates call ag-gag laws. Some of the laws made it a crime to take photos or videos of private farm property without the owner's permission, while others made it a crime for an employee of an animal welfare organization to lie about where they worked when they applied for a job on a farm."[5]

State statutes

Nine states in the United States have passed ag-gag laws.[6] They are Arkansas, Idaho, Iowa, Kansas, Missouri, Montana, North Carolina, North Dakota,

[2] Prygoski, A. (2015). A Brief Summary of Ag-gag Laws. *Michigan State University, Animal Legal & Historical Center.* https://www.animallaw.info/intro/ag-gag-laws.

[3] The Humane Society of the United States. *Anti-Whistleblower Ag-gag Bills Hide Factory Farming Abuses from the Public.* https://www.humanesociety.org/resources/anti-whistleblower-ag-gag-bills-hide-factory-farming-abuses-public (last visited Nov. 2019).

[4] Michigan State University. *Animal Legal & Historical Center.* https://www.animallaw.info/statutes/topic/ecoterrorism-or-agroterrorism?order=title&sort=desc (last visited Oct. 2019).

[5] PEW. (2016). *Farmers Push Back Against Animal Welfare Laws.* https://www.pewtrusts.org/en/research-and-analysis/blogs/stateline/2016/11/29/farmers-push-back-against-animal-welfare-laws.

[6] The National Agricultural Law Center held a webinar on ag-gag laws and the subsequent litigation on Nov. 13, 2019. The presentation is publicly available on their website. https://nationalaglawcenter.org/consortium/webinars/aggagoverview/ (last visited Nov. 2019).

and Utah. These laws were generally passed between 2010 and 2015. The American Society for the Prevention of Cruelty to Animals (ASPCA) keeps a full list of states that have introduced ag-gag legislation.[7] The list includes twenty-nine states. While many of these statutes did not pass and others have already been struck down by the courts, others remain or are currently in the midst of litigation (described below).

Case law

Many organizations track ag-gag laws around the country. Animal Legal Defense Fund (ALDF) is one such organization. Their map of ag-gag laws is seen in Figure 4.1. ALDF has been involved in litigation in an effort to prevent enforcement of state enacted ag-gag laws. Their first lawsuit was filed against the State of Utah in 2013. In 2014, they filed suit against the State of Idaho. In August of 2015, the U.S. District Court for the District of Idaho ruled that Idaho's law was unconstitutional. In 2016, ALDF joined others in suing the State of North Carolina over its "Anti-Sunshine Law." In July of 2017, the U.S. District Court of Utah found its ag-gag law to be unconstitutional. In 2017, the ALDF and others filed suit in Iowa against that state's ag-gag law. In January of 2018, the U.S. Court of Appeals for the Ninth Circuit issued its ruling on the appeal from Idaho. That decision is excerpted below. In 2018, ALDF and others filed suit against the State of Kansas. In 2019, the Federal District Court for the Southern District of Iowa ruled that Iowa's ag-gag law was unconstitutional. Iowa then enacted a revised ag-gag law, which is currently being challenged in court. In 2019, ALDF filed suit against the State of Arkansas.[8]

There have been two federal appellate court decisions on the merits of ag-gag laws. They are particularly relevant for understanding the types of ag-gag laws enacted and the reasons they are often being struck down. In 2017, the 10th Circuit Court of Appeals issued a decision regarding Wyoming's ag-gag law. It was different from many of the other ag-gag laws enacted, because it specifically targeted "collecting resource data" (i.e., at environmental advocates collecting samples of pollution or other environmental degradation due to agriculture). In 2018, the 9th Circuit Court of Appeals issued a decision regarding Idaho's ag-gag law. This law was more similar to the type of ag-gag laws enacted in other states, which specifically penalized documentation of animal welfare within agricultural facilities. Both statutes

---

[7] ASPCA. *What is Ag-gag Legislation?* https://www.aspca.org/animal-protection/public-policy/what-ag-gag-legislation (last visited Nov. 2019).
[8] *Ibid.*

Figure 4.1 Animal legal defense fund ag-gag map.[9]

were challenged on constitutional grounds. Both decisions are summarized below.

Western Watersheds Project v. Michael[10]

"Criminal trespass occurs when an individual enters or remains on the property of another with knowledge or subsequent notification that he is not authorized to do so." Trespass was already illegal in Wyoming when in 2015, it passed two statutes that prohibited people from entering

---

[9] Animal Legal Defense Fund. *Ag-gag Issues.* https://aldf.org/issue/ag-gag/ (last visited Oct. 2019).

[10] *Western Watersheds Project v. Michael*, 869 F.3d 1189 – Court of Appeals, 10th Circuit (2017).

"open land for the purpose of collecting resource data" without permission. "Resource data" is "data relating to land or land use," including anything pertaining to "air, water, soil, conservation, habitat, vegetation or animal species." "Collect" means: "(1) taking a 'sample of material' or a 'photograph,' or 'otherwise preserving information in any form' that is (2) 'submitted or intended to be submitted to any agency of the state or federal government.'" Data collected in any way violating this statute could not be used and the government was required to delete any information collected unlawfully and could not rely on it in making any official decisions.

The plaintiffs argued that the statutes were contrary to Constitution's First and Fourteenth Amendments, Free Speech and Equal Protection (among others). The trial court determined that free speech, petition, and equal protection claims were valid. Wyoming then modified the statutes.

The new statutes eliminated the term "open lands" and prohibited people from (1) entering private land "for the purpose of collecting resource data"; (2) entering private land to "collect resource data"; or (3) "crosses private land to access adjacent or proximate land where he collects resource data." Collect was redefined as (1) "to take a sample of material" or "acquire, gather, photograph or otherwise preserve information in any form"; and (2) "recording ... a legal description or geographical coordinates of the location of the collection."

The trial court then ruled for Wyoming and the plaintiffs filed an appeal with the appellate court. Though some of the statutes apply to conduct that occurs on private land and is, therefore, not subject to constitutional protection, the appellate court determined it was mistaken as to the provision about crossing private land to collect data on adjacent land, which could apply to activity on public property.

Because Wyoming already prohibits trespass, this statute has the effect of increasing "a pre-existing penalty for trespassing if an individual subsequently collects resource data from public land." This results in differential treatment. The appellate court said, "the question is not whether trespassing is protected conduct, but whether the act of collecting resource data on public lands qualifies as protected speech."

The statutes penalized activity on public land, "so long as an individual also records where such data was gathered." These prohibited activities could include "collecting water samples, taking handwritten notes about habitat conditions, making an audio recording of one's observation of vegetation, or photographing animals."

The appellate court held that collecting "resource data" on public land is protected speech in that it impacts the creation of speech common in public policy advocacy and development. It stated that the United States

"Supreme Court has explained that 'the creation and dissemination of information are speech within the meaning of the First Amendment.'"[11]

Animal Legal Defense Fund v. Wasden[12]

In 2012, an undercover activist obtained employment at a farm and surreptitiously filmed inhumane treatment of the animals. In 2014, Idaho passed a statute prohibiting "interference with agricultural production." The law made it a crime to (1) enter an "agricultural facility by force, threat, misrepresentation or trespass"; (2) obtain records by the same means; (3) obtain employment through misrepresentation with the intent to harm the farm, including economically; or (4) record conduct at a farm.[13]

The appellate court considered the underlying motivations behind the state's enactment of the law. "The bill was drafted by the Idaho Dairymen's Association, a trade organization representing Idaho's dairy industry." One of the stated goals was to protect the farming industry from "undercover investigators who expose the industry to the 'court of public opinion,' which destroys farmers' reputations, results in death threats, and causes loss of customers."

Plaintiffs asserted the statute violated the First and Fourteenth Amendments of the U.S. Constitution, similarly to the *Western Watersheds* case summarized above. The trial court ruled in their favor. The appellate court considered the matter in three sections, as itemized below.

I. The misrepresentation clauses

The appellate court explained the issue as whether or not "the misrepresentations prohibited in the Idaho statute constitute[d] speech protected by the First Amendment." It then went on to analyze the issue based on the United States Supreme Court case of *Alvarez*. In that case, the Supreme Court determined that false statements can be protected speech. However, "false speech may be criminalized if made 'for the purpose of material gain' or 'material advantage,' or if such speech inflicts a 'legally cognizable harm.'"

A. Entry by misrepresentation

This section of the law prohibits entry into a farm "by force, threat, misrepresentation or trespass." This language is problematic because under Alvarez, "a false statement made in order to access an agricultural production facility — cannot on its face be characterized as 'made to effect

---

[11] All quotes within this summary are *ibid.*
[12] *Animal Legal Defense Fund v. Wasden*, 878 F. 3d 1184 – Court of Appeals, 9th Circuit (2018).
[13] Idaho Code § 18-7042(1)(a)-(d).

a fraud or secure moneys or other valuable considerations.'" If a statute seeks to prohibit false statements regardless of whether the speech was intended to profit is suspect because it may impact innocent conduct. It is also overly broad and appears to target the speech of reporters. The appellate court determined that the misrepresentation clause regulated protected speech and must be constrained to "achieve a compelling government interest."

The appellate court said, "we are also unsettled by the sheer breadth of this subsection given the definitions of 'agricultural production facility' and 'agricultural production.'" The appellate court reasoned that the law could apply to grocery stores or plant nurseries or even hardware stores. "The subsection's reach is particularly worrisome because many of the covered entities are, unlike large-scale dairy facilities, places of business that are open to the public."

The appellate court also stated that the law was not the "least restrictive means among available, effective alternatives." It went on to state "we do not ignore that a vocal number of supporters were less concerned with the protection of property than they were about protecting a target group from critical speech."

The appellate court concluded that these concerns could be remedied by the removal of the word "misrepresentation" from the statute. In Idaho, the court may amend a law where the wording is unconstitutional and also not essential to the application of the law. "Because the proscription on misrepresentations is neither integral nor indispensable to the subsection's goal of protecting property rights, the offending term 'misrepresentation' should be stricken, leaving the remainder of the subsection intact."

B. Obtaining records by misrepresentation

This section of the law "criminalizes obtaining records of an agricultural production facility by misrepresentation." The appellate court reasoned that this portion of the law dealt with a specific harm and, therefore, was not prohibited under Alvarez. "Alvarez highlights that a false statement made in association with a legally cognizable harm or for the purpose of material gain is not protected. Unlike false statements made to enter property, false statements made to actually acquire agricultural production facility records inflict a property harm upon the owner, and may also bestow a material gain on the acquirer."

The appellate court concluded that this statute impacted lawful property rights. It also determined that Idaho lawmakers had sufficiently demonstrated how the conduct implicated in this section of the law can harm farmers. "Although some legislators wanted to silence investigative journalists reporting on the agricultural industry, the full legislative history shows that a legitimate purpose for enacting the subsection was to prevent harm from damaged or stolen records."

It was also plausible that obtaining these records could result in profit to the person misrepresenting themselves as the records could contain proprietary information or trade secrets. "Acquiring records by misrepresentation results in something definitively more than does entry onto land — it wreaks actual and potential harm on a facility and bestows material gain on the fibber. So unlike subsection (a), subsection (b) does not regulate constitutionally protected speech, and does not run afoul of the First Amendment."

The appellate court also determined that this section of the statute did not violate the Equal Protection Clause. "Legislation is generally presumed to be valid and will be sustained under the Equal Protection Clause 'if the classification drawn by the statute is rationally related to a legitimate state interest.'" The appellate court found that the "overall purpose" of the law was to protect farms from harm. "Idaho's desire to protect against harm relating to an agricultural production facility's most sensitive information — affecting both property rights and privacy interests — is a legitimate government interest…. Subsection (b) does not offend the Equal Protection Clause because it does not rest exclusively on an 'irrational prejudice' against journalists and activists."

C. Obtaining employment by misrepresentation

This section of the law prohibited "obtaining employment with an agricultural production facility by … misrepresentation with the intent to cause economic or other injury." The appellate court stated that this section of the law strictly adhered to the Alvarez decision. "Alvarez explicitly stated that 'where false claims are made to effect a fraud or secure moneys or other valuable considerations, say offers of employment, it is well established that the Government may restrict speech without affronting the First Amendment.'" The appellate court thought it was very significant that the Supreme Court specifically included offers of employment.

The appellate court also found that this section of the law did not violate the Equal Protection Clause, because it serves a "legitimate governmental purpose." This is because employees have different levels of access to the facility and its records, which may be confidential.

II. The recordings clause

Finally, the appellate court considered the provision that prohibited video or audio recordings of animal abuse or other inappropriate conduct. "The Recordings Clause regulates speech protected by the First Amendment and is a classic example of a content-based restriction that cannot survive strict scrutiny."

The appellate court confirms that recordings are protected speech, just like a book. "It defies common sense to disaggregate the creation of the video from the video or audio recording itself. The act of recording is itself an inherently expressive activity; decisions about content, composition, lighting, volume, and angles, among others, are expressive in the same way as the written word or a musical score…. Because the recording process is itself expressive and is "inextricably intertwined" with the resulting recording, the creation of audiovisual recordings is speech entitled to First Amendment protection as purely expressive activity."

This section of the law also targets specific content, which makes it prohibited regulation of "content-based speech." It intended to prohibit speech targets at a particular subject matter. "A regulation is content-based when it draws a distinction 'on its face' regarding the message the speaker conveys or 'when the purpose and justification for the law are content based.'" The section does both. Therefore, it "prohibited public discussion of an entire topic." Further, because the section prohibits filming of the operations, it is specifically targeted at the content of the recording. "In other words, only by viewing the recording can the Idaho authorities make a determination about criminal liability. Here, the statute depends not just on 'where they say' the message but also — critically — 'on what they say.'"

Because the law is content based, it must be "necessary to serve a compelling state interest" and "is narrowly drawn to achieve that end." The appellate court did not think it could meet the latter part of the test, no matter if it could meet the former. "We are left to conclude that Idaho is singling out for suppression one mode of speech — audio and video recordings of agricultural operations — to keep controversy and suspect practices out of the public eye."

"For these reasons, the Recordings Clause cannot survive First Amendment scrutiny and is therefore unconstitutional. In light of this result, we need not analyze the Recordings Clause under the Equal Protection Clause."[14]

Discussion

1. What do you think the purpose of ag-gag laws are? Are they ever justified?
2. Does the state have an obligation to balance the business interests of agriculture and the public's interest in a healthy environment or humane treatment of animals? How can that balance be reached?
3. Is there an ag-gag law in your state? If yes, read it. Do you agree with it or not? Why?

---

[14] All quotes within this summary are from the *Wasden* case, cited above.

4. Do you agree with the courts that these statutes violate free speech? Why? Under what circumstances should speech be limited?
5. Should activists be allowed to go "undercover" to document violations of the law even if that means they might misrepresent themselves or their purpose? Why?
6. How might ag-gag laws impact transparency within our food system? What might the ramifications be for consumers?

part two

Food safety and health

chapter five

Foodborne disease outbreak investigation: Surveillance and responses in the United States, Canada, the European Union, and globally

Justin Falardeau, Karen Fong, and Siyun Wang

Contents

EDITORS' NOTE: TRACEABILITY, TRANSPARENCY, AND OUTBREAK INVESTIGATION

Three food safety scientists provide resolute context to the importance of foodborne disease outbreak investigations at local and multinational levels. The authors describe how foodborne outbreak investigations are conducted, the responsibilities for the investigation, and management of foodborne outbreaks and provide case studies on landmark foodborne outbreaks during the late 2010s – including the Yuma, Arizona romaine lettuce outbreak of *Escherichia coli* O157:H7 in April 2018 that sickened at least 210 people across 36 states, hospitalized 96 people, and killed at least 5 people.

While Chapter 4 discusses the financial and legal implications involved in food safety failures, this chapter explains the essential role of outbreak investigators to seek and mitigate food safety and public health threats. Since the passage of the U.S. Food and Drug Administration (FDA) Food Safety Modernization Act (FSMA) of 2011, nations around the world have focused on improving food safety standards to promote international trade and overall health, but certain foodborne hazards continue to threaten the safety of the food supply. To help improve transparency and decrease the time it takes to conduct a thorough investigation, a greater focus on traceability and enhanced recordkeeping became a priority among government and industry officials.

In October 2018, a group of nonprofit organizations, the Center for Food Safety and Center for Environmental Health, successfully sued the FDA after the agency missed key congressional deadlines to meet statutory FSMA deadlines pertaining to high-risk foods. In September 2020, the U.S. FDA introduced a public comment period for FSMA Section 204(d)(2)(A) that seeks to require nonexempt food businesses to designate "high-risk foods" for which additional recordkeeping requirements are appropriate and necessary to protect the public health. These food items will require firms to enhance traceability capacity to help outbreak investigators more rapidly find possible root causes of contamination events. This rule will create a "Food Traceability List," and also define the terms "Critical Tracking Events" and "Key Data Elements" to promote interoperability of traceability systems.

Introduction

Foodborne illness is an ongoing global issue, with the World Health Organization (WHO) estimating an annual occurrence of 600 million cases of foodborne illness and 420 million related deaths (WHO, 2015). In the United States, the Centers for Disease Control and Prevention (CDC) estimates 48 million foodborne illnesses and 3,000 deaths annually (CDC, 2018a). During foodborne outbreaks, rapid detection and response are essential to minimize the number of cases, and to develop strategies to prevent similar outbreaks from occurring in the future (Biggerstaff, 2015).

As food systems become more globalized, and distribution chains become longer and more complex, the need for efficient outbreak detection and response is especially important. Effective strategies for foodborne outbreak investigation involve four main components: (i) public health surveillance to quickly detect outbreaks, allowing for a rapid and efficient response; (ii) epidemiological expertise to identify the causative agent; (iii) supply chain traceability to identify the root source of contamination; and (iv) infrastructure for rapid control measures to remove the affected food from the market.

How are foodborne disease outbreaks investigated?

Detecting and investigating foodborne outbreaks is a multistep process. The specifics of each step might differ between outbreaks, but the basic principles remain the same. For the sake of explanation, these steps have been outlined linearly, but it is important to recognize that several steps might be happening concurrently or in a different order. In this summary, we used the seven steps described by the U.S. CDC as a backbone (CDC, 2018e) with additional information from other sources such as the WHO (WHO, 2008) and the International Association for Food Protection (Todd, 2011).

Step 1: Detect a possible outbreak

The first step in stopping an outbreak is to detect the outbreak. An outbreak is defined as two or more cases of a similar foodborne disease associated with the consumption of a common food item (Kearney, 2018). The earlier an outbreak is detected, the easier it will be to minimize the negative consequences (Marvin et al., 2009).

Outbreaks can be detected through three possible methods: Formal reports, informal reports, or public health surveillance. Formal and informal reports occur when public health authorities are contacted by health professionals or the public, respectively. This might occur when a cluster of illnesses erupt on a small scale, such as at a public event (e.g., wedding), or a higher number of people than normally expected show up to the hospital with similar symptoms. Additionally, in the United States and Canada, several foodborne pathogens, such as *Salmonella* and Shiga toxin-producing *Escherichia coli* (e.g., *E. coli* O157:H7) are notifiable diseases, meaning that doctors are encouraged to report any suspected occurrence to regional or national health authorities (CDC, 2019f; PHAC, 2020). The health authorities are then able to monitor this data for for suspicious variations through public health surveillance.

Public health surveillance involves the collection, analysis, and interpretation of public health data (WHO, 2008). The goal of public health surveillance is to identify disease clusters, which are defined as a higher-than-usual number of cases of the same illness within a particular time and place (Williams et al., 2019). Through the monitoring of this occurrence data, sharp increases in the occurrence of a particular pathogen within a specific time, area, or particular subpopulation may indicate the occurrence of an outbreak and should warrant further investigation. Public health surveillance is especially important to detect widespread (e.g., national) outbreaks missed by local and regional health authorities (WHO, 2008). In the United States, the Foodborne Diseases Active Surveillance Network (FoodNet) has been tracking trends in foodborne illness since 1996 (Henao et al., 2015).

In cases where the germ is isolated and identified, further genetic subtyping of the causative pathogen may be conducted, such as pulsed-field gel electrophoresis (PFGE), which produces a DNA fingerprint, or whole genome sequencing (WGS). These genetic data can be uploaded from local laboratories to the CDC's PulseNet system, allowing them to be compared with those of other patients and helping to identify unrecognized clusters or ongoing outbreaks (CDC, 2016b).

Step 2: Define and find cases

Once an outbreak has been detected, the next step is to establish a case definition. A case definition allows investigators to get a clear sense of who is, and who is not, part of the investigated outbreak, and should be sensitive enough to include all cases, but specific enough to exclude all those who are not part of the outbreak, even if the patients show similar symptoms (WHO, 2008). Case definitions should include the following four pieces of information:

1. Features of the illness, including the symptoms, pathogen, and genetic subtype (if known).
2. Time frame of the expected occurrence of illnesses.
3. Geographical range of expected illness occurrence.
4. Characteristics of persons likely to have contracted the illness (e.g., sex, nationality, and age).

The case definition may change throughout the investigation. A loose definition may be used early, with a more specific definition developed as more evidence is collected (Kearney, 2018). Multiple case definitions may be used simultaneously, potentially grouping cases into three categories: confirmed, probable, and possible. Confirmed cases are those which have a proven, laboratory-based result. Probable cases are those that show symptoms of the disease but have not yet been confirmed by laboratory test. Finally, possible cases are those that show some, but not all, or reduced clinical symptoms (Williams et al., 2019).

Once the case definitions have been clarified, investigators should begin tracking down additional unreported cases. This is important since usually only a small fraction of total cases are ever reported to health agencies (Scallan et al., 2011).

Depending on the outbreak, a variety of methods may be used to identify additional cases. If the outbreak is associated with an identifiable group, such as attendees of a particular social event, or students in a specific school, investigators may contact all individuals directly to inquire about illness or symptoms. If the DNA fingerprint and/or WGS data is known, investigators can look through the PulseNet database for illnesses associated with the same strain. This is especially useful for multijurisdictional outbreaks that evolve over a long period of time (WHO, 2008). If the symptoms associated with the case definition are clear, investigators can contact local hospitals and physicians. Finally, the investigators can alert the public directly through the use of the media. This provides the additional benefit that the public can be warned away from consuming or purchasing an implicated food source, while simultaneously searching for other individuals who may be part of the outbreak.

As cases are identified, they are tracked in both time and space. An epidemic curve, (or epi curve; Figure 5.1), is a histogram that tracks the number of illnesses over time. Using this information, investigators can determine the period of time over which people were exposed to the pathogen, the incubation period, and how interventions have affected the case rate. Maps may also be used to investigate geographical patterns. By looking at how cases cluster geographically, a central location may be identified as the source of the occurring outbreak, such as a particular grocery store.

Step 3: Generate hypotheses

The next step is to generate hypotheses about the likely source of the outbreak. In foodborne outbreaks, this means determining the contaminated food product. Interviews of infected individuals are often the primary source that investigators utilize in order to start narrowing down the source of the outbreak. In these interviews, investigators will question each case patient on what foods they have eaten in the past days or weeks. These interviews are driven by the information gathered during steps 1 and 2 and will focus on specific foods or other sources (i.e., restaurants)

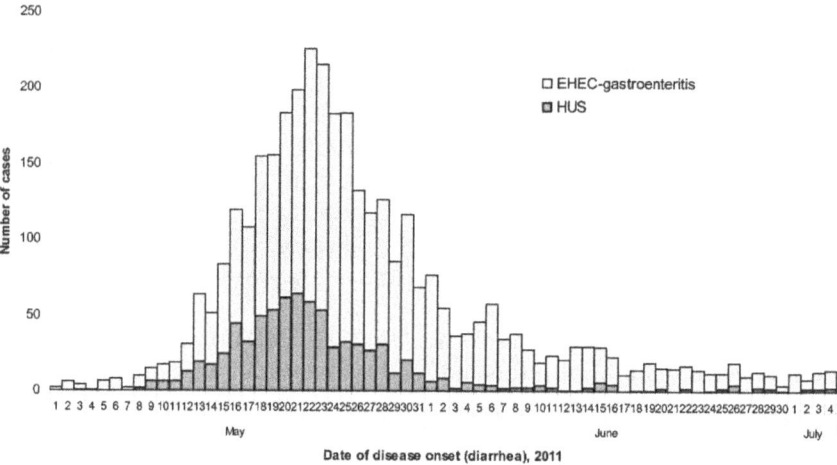

Figure 5.1 Epidemic curve of the fenugreek *E. coli* O104:H4 outbreak, 2011. (Copyright © Werber et al., 2012; Published by MioMed Central Ltd. under the Creative Commons License (http://creativecommons.org/licenses/by/2.0), which permits unrestricted use, distribution, and reproduction in any medium, provided the original work is properly cited.)

if that information is already available. Additionally, many foodborne pathogens are commonly associated with certain foods, which may also influence the focus of the interview.

In the cases where no preliminary ideas about possible sources exist, investigators may question ill people using a "shotgun questionnaire," which is a standardized questionnaire that includes a mixture of open-ended questions about dietary and food-purchasing habits, as well as questions about a long list of food items (Williams et al., 2019). Based on the results of these interviews and questionnaires, investigators may begin generating hypotheses based on common food exposures across all or most of the infected individuals.

Generating hypotheses may be difficult as people often have difficulty remembering what they have eaten over the past month, weeks, or even days (Seitzinger et al., 2019). Furthermore, the outbreak source may be a single ingredient in a multi-ingredient item, such as eggs or flour. Therefore, it may take multiple interviews with multiple individuals before a suitable hypothesis can be generated.

Step 4: Testing of hypotheses

Once hypotheses have been generated, investigators must test them in order to determine if the identified food or source is correct. Evidence is founded on three "pillars" of information; however, it is often impossible to confirm all three. First, the investigators should confirm that there is a correlation between exposure to the identified food and becoming ill. Second, traceback should point to a common source of the identified food for the majority, if not all, of the cases, and finally, laboratory tests should confirm the presence of the affecting agent at the source and in the identified foods (Williams et al., 2019).

Confirmation of the source food can be conducted using case-control studies or cohort studies. The purpose of these two study types is to determine if ill persons were more likely to have consumed the hypothesized food. Cohort studies can be used when an outbreak is associated with a small, identifiable event, such as a gala reception. By interviewing each attendee about which foods they ate, the proportion of ill people who ate a particular food, known as the attack rate, can be used to look for likely culprits. The relative risk of each food item can also be calculated as the ratio of attack rate of those who consumed the food to those who did not consume the particular food. High relative risks increase the evidence against that particular food. Case-control studies, on the other hand, can be used when there is no easily identified "cohort" associated with the outbreak. The group of cases can be compared to a comparable group of controls, such as nonill individuals, to assess exposure to foods. Using case-control studies can determine the odds that exposure to a particular food resulted in illness, helping investigators confirm likely candidates.

Many foodborne outbreaks cannot be linked to one particular location or event, however, and so the common source must be identified along the supply chain. Traceback can be used to identify the source of an ingredient when illnesses cannot be associated with a particular source such as restaurant or grocery store. By tracing back through the respective supply chains, a common source of convergence can be identified. This may be a particular food plant that produced one of the ingredients, or it may link all the way back to the farm of origin.

Once the likely source has been identified, laboratory testing can be ideally conducted to confirm the presence of the suspected pathogen. In a perfect world, a bacterium that possesses a matching DNA fingerprint from the case patients will be isolated from an unopened food package, or from within the production facility. For a myriad of reasons, however, this might not be possible. For example, the food items in question may no longer be available for testing due to a short shelf-life, or detection/isolation methods may not currently exist for the pathogen in question (Williams et al., 2019).

Step 5: Solve the point of contamination

Once the suspected location or source of the illness has been determined, environmental testing can be used to determine the point of contamination (i.e., how the foodborne pathogen found its way into the food). If the processing facility responsible for preparing the food is identified, investigators may interview employees about production procedures or review the cleaning and sanitation protocols. Investigators may also review past inspection reports to identify areas of concern that had been flagged previously. Investigators may also take environmental and/or food samples from the facility to test for the presence of the affecting pathogen. This may help pinpoint the exact point of contamination, and may also identify where corrective action needs to be taken.

Step 6: Control the outbreak

If and when the source of the outbreak has been confirmed, measures should be put in place to stop the outbreak and to prevent a similar outbreak from reoccurring. These typically include the removal of implicated foods from the market (i.e., a product recall), modifying the production practices of the processing facility, and in extreme cases, closing down the implicated food premises. Additionally, public health officials should communicate with the public any necessary information to prevent further illness, such as throwing away recalled foods, and warning them about foods to avoid that may still be in circulation.

Step 7: Decide the outbreak is over

An outbreak is considered over once the number of observed cases falls back to the regular baseline. Based on the epidemic curve, outbreak investigators can monitor the effects that their interventions had. Investigators should continue to monitor for cases for a few weeks before confirming the outbreak to be completely over. If the number of illnesses does not decline, or continues to rise, it suggests that the true source of the outbreak has not been completely controlled, and the investigation should be continued.

Responsibilities for the investigation and management of foodborne outbreaks

Food safety and a timely response to foodborne outbreaks is an important responsibility of governments, whether local, national, or international. To that end, several agencies and programs have been set up whose primary responsibility is to conduct food safety surveillance and analysis, and to detect and respond to foodborne outbreaks. Here, we have summarized some of those agencies and programs that exist in the United States, Canada, the European Union, and globally.

United States

In the United States, local health authorities are responsible for local surveillance and investigation of foodborne illnesses and outbreaks. They collect complaints about illnesses or food-service operations, and interview ill persons for information gathering during outbreak investigations. Additionally, each state has a variety of state-wide health agencies who provide support to local health authorities through epidemiological, environmental, and laboratory assistance. These state-level health authorities also coordinate the investigation of multijurisdictional outbreaks, or those related to commercial food products within their state borders (Council to Improve Foodborne Outbreak Response, 2014).

At the federal level, three primary agencies are responsible for the safety of the food supply: The United States Centers for Disease Control and Prevention (CDC), the United States Food and Drug Administration (FDA), and the United States Department of Agriculture (USDA). Through myriad programs, these three agencies work individually and in tandem to detect, investigate, and prevent foodborne illness outbreaks.

National surveillance of foodborne disease outbreaks falls on the shoulders of the CDC. Through the National Surveillance of Bacterial Foodborne Illnesses, the CDC collects and analyzes data on nationally notifiable enteric bacterial pathogens provided by local and state health authorities (CDC, 2020). When bacterial strains are isolated by the local and state health laboratories, the genetic subtyping data of these pathogens are collected and stored by the CDC through PulseNet (CDC, 2019a). These genetic subtyping data allow the CDC to more quickly identify outbreaks, especially those related to multistate clusters that might not be recognized by local and state health authorities (CDC, 2016a). Finally, the Foodborne Diseases Active Surveillance Network (FoodNet) collects laboratory-based data about foodborne illness at various sites around the United States. This active surveillance, which covers 15 percent of the United States total population, allows the CDC to determine and monitor trends of foodborne illness across the United States (CDC, 2019c).

During large and/or multistate outbreaks, the CDC provides increased laboratory capacity, acts as a central data system for collection and analysis of ongoing outbreak data, and is able to provide coordination and assistance to state and local health agencies during the investigation and response. Response to multistate outbreaks is coordinated through the Foodborne Outbreak Response Team (FORT). FORT coordinates with local, state and other federal agencies during the response and investigations, and communicates warnings to the public during these outbreaks (CDC, 2018c).

The CDC is also mandated to improve training, prevention, detection, and investigation strategies related to foodborne outbreaks. To this end, several programs exist within the CDC. In order to provide a deeper understanding of the common causes of foodborne illness outbreaks, the Foodborne Disease Outbreak Surveillance System (FDOSS) collects and analyzes information from local and state health department investigations of foodborne illness (CDC, 2018b). During and after outbreak investigations, the Environmental Health Specialists Network (EHS-Net) conducts environmental assessments with the goal of identifying contributing factors related to food contamination, and to use this data to prevent future outbreaks (CDC, 2019b). Finally, the CDC has two programs to develop tools and resources, and to provide training and education for local and state health authorities. These are the Integrated Food Safety Centers of Excellence program (CDC, 2019e), which partners with academic institutions in five states as a resource for public health officials; and the Foodborne Diseases Centers for Outbreak Response Enhancement (FoodCORE) program (CDC, 2019d), which focuses on developing and improving methods for detection, investigation, and response to enteric foodborne outbreaks. All of these programs are also available for assistance during active outbreak investigations.

Under the Food Safety Modernization Act, the FDA was named as the lead agency for food safety in the United States. The FDA regulates the safety of all foods not covered by the USDA (i.e., meat, poultry, and pasteurized egg products), and provides oversight for the safety of imported foods (Council to Improve Foodborne Outbreak Response, 2014). Foodborne illness outbreak surveillance, response, and postresponse is handled by the Coordinated Outbreak Response and Evaluation Network (CORE). The CORE team conducts surveillance in parallel with the CDC and has three response teams responsible to control and stop the outbreak. These response teams are able to conduct traceback investigations along distribution lines, and work with state and local regulators and health departments to remove contaminated products from circulation (FDA, 2020a). The CORE network also contains a Post-Response Team who investigates factors that led to the outbreak in order to develop prevention strategies moving forward. In select states, the FDA also has Rapid Response Teams (RRTs) which partner with state authorities during food outbreak emergencies (FDA, 2019).

The USDA has a similar role to the FDA, but is responsible for meat, poultry, and pasteurized egg products. Oversight, regulation, and response to foodborne illness is conducted through the Food Safety Inspection Service (FSIS; USDA, 2019). The FDA and USDA also co-lead the Food Emergency Response Network (FERN), which was created in response to the threat of foodborne terrorist attacks, but has also been important in large-scale multistate outbreaks (United Federal State & Local Laboratories for Food Emergency Response, 2019).

Canada

In Canada, foodborne outbreaks associated with a single jurisdiction are handled by local/regional or provincial/territorial health officials based on protocols that differ between regions and provinces and territories. These local/regional or provincial/territorial health officials may request assistance from federal authorities as necessary (PHAC, 2017).

Nationally, food safety in Canada is the responsibility of three federal authorities: the Public Health Agency of Canada (PHAC), the Canadian Food Inspection Agency (CFIA), and Health Canada. Public health surveillance, coordination of outbreak response during multijurisdictional outbreaks, and communication with the public during foodborne outbreaks is the responsibility of PHAC. The CFIA provides enforcement of food safety in Canada and is tasked with conducting food safety investigations and recalling contaminated food products. Finally, Health Canada is responsible for setting food safety standards and regulations in Canada, and is not directly involved in outbreak investigations; however, they may conduct health risk assessments of suspected foods at the direction of the CFIA (PHAC, 2017).

The food safety and surveillance mandates of PHAC are shared by two divisions. The forward-facing Centre for Food-borne, Environmental and Zoonotic Infectious Diseases (CFEZID) is responsible for public surveillance of enteric illnesses and management of outbreak response (PHAC, 2013a). On the other side, the Enteric Diseases division of the National Microbiology Laboratory (NML) provides laboratory-based surveillance and assistance (PHAC, 2019a).

National surveillance for foodborne illness outbreaks in Canada is conducted through the National Enteric Surveillance Program (NESP), which is jointly administered by CFEZID and the NML. The NESP collects weekly reports of confirmed enteric diseases from provincial public health laboratories through the Canadian Public Health Laboratory Network, which allows for the detection of emerging trends, and identification and response to emerging clusters of foodborne illness (PHAC, 2018).

The PHAC also runs two other food safety surveillance programs modeled after their counterparts in the United States. FoodNet Canada performs comprehensive enteric illness surveillance at three sentinel sites across Canada. This collected surveillance data is then analyzed in order to better understand and to reduce the burden of foodborne illness in Canada (PHAC, 2013b). Additionally, PulseNet Canada, under the NML, fills a similar role as PulseNet in the United States. In partnership with provincial public health laboratories, PulseNet Canada is a virtual database that collects the DNA fingerprints and WGS of *E. coli*, *Salmonella*, and *Listeria monocytogenes* associated with foodborne infection. This data is then used for early identification and efficient investigations of foodborne disease outbreaks (PHAC, 2019b).

The CFIA is the enforcement arm of Canada's food safety system. When a suspected outbreak is identified, the CFIA is responsible for conducting the food safety investigation in order to determine the source of the foodborne illnesses. This involves tracing suspected foods through the production and distribution chains in order to pinpoint the particular source or ingredient. Once a suspected source has been identified, the CFIA will conduct on-site investigations in order to determine the root cause of the problem and identify necessary corrective actions. The CFIA may also request a health risk assessment from Health Canada. If sufficient risk associated with the food product has been identified, the CFIA will implement and oversee (or enforce as necessary) the effective removal of the contaminated product from the marketplace. The CFIA will also communicate with the public to inform them of any recalled product that may be present in their homes (CFIA, 2019).

European Union

Within the European Union (EU) and European Economic Area (EEA), Member States are each responsible for food safety surveillance and outbreak response within their respective borders. Multinational oversight, however, is provided by two agencies of the EU: The European Centre for Disease Prevention and Control (ECDC) and the European Food Safety Authority (EFSA). The ECDC, through the Food- and Waterborne Diseases and Zoonoses Program provides support to Member States for public health surveillance and during response to multinational foodborne outbreaks (ECDC, 2020b). The EFSA, on the other hand, preforms ongoing risk assessments of foodborne illness in the EU/EEA through the annual collection of data on foodborne illnesses and outbreaks from Member States (European Food Safety Authority and European Centre For Disease Prevention and Control, 2019).

General surveillance for foodborne illness across the EU/EEA is conducted through the Epidemic Intelligence Information System for Food- and Waterborne Diseases and Zoonoses (EPIS-FWD).This program, run jointly by the ECDC and EFSA, collects genetic subtyping data of isolates of common foodborne pathogens from EU/EES Member States. The resulting data is accessible to fifty-two countries worldwide, making it an important tool in global food safety (Gossner, 2016). Additionally, the European Food- and Waterborne Diseases and Zoonoses Network (FWD-Net) of the ECDC monitors foodborne illness trends throughout Member States, and helps to detect and monitor multinational foodborne illness outbreaks (ECDC, 2020a).

International

Globally, several programs and networks aim to support and improve international efforts to detect and prevent foodborne outbreaks. The Global Foodborne Infections Network (GFN) is an international network under the WHO whose mandate is to improve surveillance capacity globally in order to foster collaboration for food safety. The GFN provides training and support for those countries still developing their foodborne surveillance programs (WHO, 2020b).

Another program managed by the WHO, in partnership with the Food and Agriculture Organization of the United Nations (FAO), is the International Food Safety Authorities Network (INFOSAN), which represents 190 national food safety authorities globally. Their goal is to share communication about ongoing food safety risks and prevent the spread of contaminated foods between countries (WHO, 2019).

International laboratory-based surveillance of foodborne pathogens is conducted through PulseNet International, a global network of public health laboratories. The network, representing eighty-six countries, is made up of smaller laboratory networks from across Africa, Asia Pacific, Canada, Europe, Latin America and the Caribbean, the Middle East, and the United States. Through this network, PulseNet International is able to standardize genetic typing methods and share information in support of global food safety surveillance (Nadon et al., 2017).

Case studies

In this section, we present three cases based on previous foodborne outbreaks and include epidemiologic methods that were practiced at the time. These cases provide insights regarding how to control foodborne disease outbreaks in the future. When reading these cases, please consider answering the following questions:

1. Based on the *E. coli* O157:H7 outbreak associated with romaine lettuce, can you identify the seven steps involved in the investigation of these outbreaks?
2. Based on the *E. coli* O104 outbreak associated with fenugreek sprouts, what were the agencies involved, and can you explain the responsibilities of these agencies during the investigation?
3. What is the role of transparency in foodborne outbreak investigations?

E. coli O157:H7 outbreak associated with romaine lettuce, 2018

Shiga-toxin producing *Escherichia coli* (STEC; e.g., *E. coli* O157:H7) is a foodborne pathogen that is normally associated with beef products, but in recent years, several outbreaks have highlighted its prevalence in fresh produce items. Symptoms of STEC infection vary, but include vomiting, abdominal cramping, bloody diarrhea, and in worst-case scenarios, hemolytic uremic syndrome (HUS), a condition characterized by kidney failure, or death (Croxen et al., 2013).

The largest multistate STEC outbreak in the United States occurred in 2018, resulting in 210 illnesses and five deaths (CDC, 2018d). Initially, two infection clusters of *E. coli* O157:H7 were reported to the CDC by the New Jersey Department of Health and the Pennsylvania Department of Health. Initial reports of food exposures pointed to the fact that several case-patients had reported eating salads at a national restaurant chain prior to becoming ill. Based on these reports, the CDC developed a questionnaire to pinpoint the contaminated food. Romaine lettuce consumption the week prior to illness was reported by 85 percent of the 179 respondents; a significant increase compared to 47 percent reported in the FoodNet Population Survey (Bottichio et al., 2019).

Subsequent traceback efforts could not identify a single lot, processor, or farm as the source. Cumulatively, thirty-six fields from twenty-three farms were identified during the traceback investigation, and all contaminated romaine lettuce was traced to farms in the Yuma growing region in Arizona; however, the source of STEC was never identified.

Since romaine lettuce is not a natural reservoir of *E. coli* O157:H7, the route of contamination onto this food vehicle is still currently unknown. There are, however, plausible hypotheses. *E. coli* O157:H7 isolates with the same DNA fingerprint as the outbreak strain were detected in a 3.5-mile section of the Wellton-Mohawk irrigation canal located in the Yuma growing region that runs adjacent to the romaine lettuce farms that had been identified during the traceback investigation. This canal also runs adjacent to a cattle-rearing operation that houses approximately 105,000 heads of cattle. Contamination of the romaine lettuce could have occurred during irrigation via water contaminated by fecal runoff. Unfortunately, a direct route of transmission was never identified (FDA, 2018).

As of June 28, 2018, no new cases had been reported. Furthermore, the last shipments of romaine lettuce from the region had been harvested on April 16, 2018, meaning that the contaminated lettuce should no longer be available in the market. Therefore, this outbreak was declared to be over (CDC, 2018d).

This outbreak drove implementation of several task forces to prevent future outbreaks, such as the Leafy Greens Food Safety Task Force and the Romaine Task Force. Initiatives proposed by both these groups included increasing distance between cattle operations and romaine farms, improvements to traceability (e.g., increased information-sharing) and changes in agricultural water practices. In particular, the overall objective of the Romaine Task Force was to examine critical areas where improvements were to be implemented throughout the romaine supply chain. In collaboration with key governmental agencies in the United States, including FDA and CDC, four areas of focus were identified in which key changes were to be implemented: Science and Prevention, Traceability, Investigations/Collaboration, and Provenance Labeling. Several recommendations were put forth in these key areas, including the establishment of an industry-driven database housing genomic information of produce-associated pathogens; tighter communication between retail facilities and primary production personnel (e.g., a trace-forward approach); and greater in-depth auditing of primary production facilities (Romaine Task Force, 2019).

E. coli *O104:H4 outbreak associated with fenugreek sprouts, 2011*

A major outbreak of STEC attributed to fenugreek sprouts occurred in 2011 in Germany. In total, 3,816 individuals were infected, progressing to 845 patients presenting with hemolytic uremic syndrome, which resulted in 54 fatalities (Knödler et al., 2016). This outbreak was the largest outbreak of HUS known to occur (Figure 5.1; Werber et al., 2012).

It took a considerable amount of effort by European authorities to identify fenugreek seeds as the implicated food vehicle. When affected patients were interviewed, it quickly became obvious that people did not remember in detail what they ate weeks ago. Since fenugreek sprouts are normally considered a food ingredient and often eaten with consumption of other vegetables, the public was initially advised to avoid common salad ingredients (e.g., salad greens, tomatoes) that case patients had recalled eating, therefore highlighting common gaps in patient memory and recall information. As a result of these questionnaires, food safety authorities originally focused their investigations on these types of ingredients (Buchholz et al., 2011).

Health notices were later changed to notify the public of sprouts consumption after epidemiological investigation and traceback studies pinpointed a contaminated batch of fenugreek seeds as the most probable source of the pathogen on June 24, 2011. The imported fenugreek seeds were subsequently sprouted by various producers upon arrival at their final destination, which is known to result in the proliferation of pathogens contaminating the seed (Gault et al., 2011). Using a traceback approach, health authorities initiated a thorough investigation of the distribution and production chain. At each step of delivery and/or production, further investigation was performed to identify and account for critical seed lots, which provided a great deal of back tracing information to various producers. Eventually, lot #48088 of fenugreek seeds imported by an importer in Egypt was identified as the most probable source (EFSA, 2011).

As was similar with the romaine lettuce outbreak (FDA, 2018), this pathogen was never isolated from the seed itself. Investigation of the production facility did not reveal any environmental contamination. Some employees were found to be infected, but did not exhibit signs of illness prior the outbreak (EFSA, 2011). This may have been because the entire lot of seeds exceeded 15,000 kg, and if only a minor part had been contaminated, it would have likely been missed during bacteriological testing (Burger, 2012).

Data transparency played an important role in the investigation of this outbreak. The outbreak strain of *E. coli* was unique and had never been identified previously. This particular strain had combined attributes of two other serotypes of *E. coli*, which resulted in greater virulence and thus increased risk to public health and food supply systems. Open-source analysis of the genomic sequence of this pathogen was undertaken, facilitated by crowd-sourced analyses and open-source data release to characterize this previously unknown pathogen. Genomic information was made readily available to bioinformaticians worldwide in the public domain, which resulted in a large burst of crowd-sourced analyses on four continents. Within one week of data release, two dozen reports on the strain analysis had been generated. These reports provided information on the strain's virulence, antibiotic resistance, and its phylogenetic lineage (Rohde et al., 2011).

Listeriosis outbreak associated with polony, 2017–2018

Listeria monocytogenes is a foodborne pathogen that causes listeriosis and is commonly associated with a variety of ready-to-eat (RTE) foods, such as processed meats and soft cheeses. Most notably, the mortality rate of *L. monocytogenes* can be as high as 30 percent in some cases. Although symptoms may range from generally mild (e.g., vomiting, diarrhea), serious side effects include preterm abortion in pregnant women. Some subpopulations are more susceptible to listeriosis than others; these include the elderly, young (i.e., neonates), immune-compromised, and pregnant women (Ramaswamy et al., 2007).

The largest ever outbreak of *L. monocytogenes* was caused by a contaminated processed meat product ("polony", similar to a bologna sausage) produced by Enterprise Foods in South Africa (Figure 5.2). In total, 1,060 cases were reported between the period of January 1, 2017 to July 17, 2018, with 216 deaths. In line with its microbiological features, pregnant women and HIV-infected individuals were disproportionately affected. Given that South Africa is a country with a high prevalence of HIV-infected individuals and high fertility rate, the increased incidence of susceptible subgroups greatly exacerbated the case count. Such information on these features is essential since it allows for more accurate

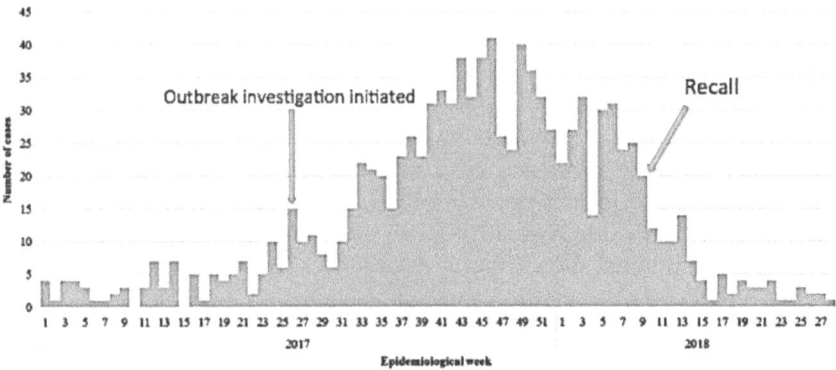

Figure 5.2 Epi-curve of the South African listeriosis outbreak. (Data adapted from Smith et al., 2019 [https://www.ncbi.nlm.nih.gov/pmc/articles/PMC6653791/] Copyright © Anthony M. Smith et al. 2019; Published by Mary Ann Liebert, Inc. under the Creative Commons License [http://creativecommons.org/licenses/by/4.0], which permits unrestricted use, distribution, and reproduction in any medium, provided the original work is properly cited.)

estimations of disease transmission and risk within certain populations (Smith et al., 2019).

Health authorities initiated the collection of food items from the households of sick patients and were able to isolate *L. monocytogenes* from food samples (including polony), thus initiating a traceback investigation. Investigations were centered around a suspected meat production facility and included food and environmental testing of food-contact surfaces, nonfood contact surfaces, and chilling brine. Ultimately, *L. monocytogenes* was isolated from unopened polony loaves in the production facility and also in the retail market (Thomas et al., 2020).

This outbreak is a good example of the importance of a clear case definition. At the onset of the outbreak investigation, all cases of *L. monocytogenes* infection were considered as part of the outbreak. Through the use of WGS, 91 percent of the 636 analyzed clinical isolates of *L. monocytogenes* believed to be part of the outbreak were classified as genetic subtype ST6, suggesting ST6 to be the causative strain. The case definition was then refined to include only those isolates identified as genetic subtype ST6. This was further validated when *L. monocytogenes* of genetic subtype ST6 was isolated from the Enterprise Foods production environment where the polony was processed (Smith et al., 2019).

Prior to this outbreak, there was a large lack of data related to the prevalence and epidemiology of *L. monocytogenes* in the country. In particular, lack of regulation, lack of food policy, and lack of transparency from large-scale food manufacturers played a role in the inability of health officials to pinpoint the source of the outbreak for more than a year (Hunter-Adams et al., 2018). Once identified, however, an efficient outbreak response and rapid information sharing between various governmental agencies, healthcare facilities, and private laboratories served to curb secondary cases, which was instrumental especially considering the limited capacity and resources available in South Africa. As a result of this outbreak, listeriosis is now classified as a mandatory notifiable medical condition in South Africa, thus expanding oversight of the nation's surveillance systems. Additionally, ready-to-eat meat processors are now legally required to be certified through the Hazard Analysis and Critical Control Point System, a system designed to minimize risk in food production facilities (Thomas et al., 2020).

What can we learn from nonfoodborne outbreaks? Lessons from the COVID-19 pandemic

In December 2019, the city of Wuhan in Hubei province, China, became the epicenter of an outbreak of unknown cause. The Health Commission of Hubei Province, China, initially announced an infection cluster of unexplained cases of pneumonia on December 31, 2019. Patients presented with symptoms such as impaired lung capacity, low white blood cell and platelet counts, hypoxemia, and abnormal liver and renal function. Traceback investigations pinpointed geographic associations to the Huanan seafood market, which sold freshly slaughtered game animals (Chan et al., 2020).

On January 7, 2020, a novel coronavirus (SARS-CoV-2) was isolated from the respiratory tract of a patient and later termed Coronavirus Disease 2019 (COVID-19) by the WHO. Despite extensive efforts by Chinese health authorities to control the spread of the disease, COVID-19 has since spread to other countries, including Japan, North/South Korea, Italy, Iran, the United States, and Canada. As of March 12, 2020, the number of infections worldwide has now exceeded 125,000 with 4,613 deaths (WHO, 2020a).

In its early stages, it may not be known if an outbreak caused by a novel, emerging pathogen is associated with a food vehicle (i.e., is a foodborne outbreak). Failure to discern the difference between a vector of foodborne origin or otherwise complicates methods for control, prevention strategies, and oversight regarding various agencies involved in outbreak response. Although some misconception exists among the public, an association between food and transmission of COVID-19 has not been identified (FDA, 2020b). This means the epidemiology of SARS-CoV-2 does not indicate consuming, purchasing, or handling food as a risk factor for illness.

It is known that SARS-CoV-2 is a part of a larger family of viruses known as the coronaviruses that, in the past, have jumped from animal sources to humans (i.e., is of zoonotic origin). For instance, the virus that causes severe acute respiratory syndrome (SARS) is thought to have jumped to humans from civets in 2002. Additionally, the virus causing Middle East respiratory syndrome (MERS) was thought to have originated in bats. This outbreak involved an extensive, yet complicated traceback investigation, in that teams from agencies including the Chinese Centre for Disease Control and Prevention and the Chinese Academy of Sciences have, so far, yet to identify a source, despite extensive testing at the live-animal market in Wuhan where the virus was thought to have been initially transmitted to humans (Chan et al., 2020).

Extensive, rapid information sharing (via epidemiological, clinical, and laboratory data) has guided public health decision-making and led to the implementation of vast public health measures to circumvent spread of the virus. For instance, concerted efforts by Chinese authorities led to the suspension of public transportation, transportation hub closures (i.e., airports, highways, rail stations, etc.), and quarantine measures to prevent disease transmission in Wuhan. Additionally, a daily press release system has been established in China to disseminate epidemic information as necessary. Since the declaration of COVID-19 as a global pandemic by the WHO on March 11, 2020, closures of schools, universities, restaurants, and other nonessential services have been taken place in many countries to limit spread of the virus. The WHO also currently has daily releases to advise the general public of COVID-19-related information as they happen in real-time.

The international race to characterize COVID-19 and develop appropriate measures (i.e., vaccines) have also seen a wealth of information and data being shared among members of the scientific community and health authorities (e.g., the WHO). As an example, the full genetic sequence of SARS-CoV-2 has been made freely accessible through a public platform, and as a result of this information release, rapid testing methods were developed that made it possible to accurately and quickly diagnose infections (Moorthy et al., 2020). In light of the current state of emergency, scientists and health professionals around the world have come together in solidarity to dispel rumours and misinformation around its origin, and to ensure the dissemination of accurate, reliable information to fight COVID-19 (Calisher et al., 2020).

Conclusion

With global population expansion, the Agri-Food sector is expected to increase production and processing efforts to meet demand. However, increased production and processing is also expected to come with enhanced risks for food safety, demonstrated in recent years by the multitude and magnitude of foodborne outbreaks. Through a series of carefully planned and executed outbreak investigation steps: (i) sensitive surveillance measures are capable of eliciting a quick outbreak response; (ii) causative agents are now quickly identified, (iii) implicated food vehicles are routinely pinpointed; and (iv) control measures are quickly put in place. Because of the sophisticated nature of foodborne outbreak investigation, concerted efforts by a multitude of agencies both within and across country borders are thus necessary. Additionally, collaborative information sharing between governmental agencies, industry, and research facilities are essential toward the identification, characterization, and eventual control of foodborne outbreaks.

References

Biggerstaff, G. K. (2015). Improving response to foodborne disease outbreaks in the United States. *Journal of Public Health Management and Practice, 21*(4), E18–E26. https://doi.org/10.1097/PHH.0000000000000115

Bottichio, L., Keaton, A., Thomas, D., Fulton, T., Tiffany, A., Frick, A., Mattioli, M., Kahler, A., Murphy, J., Otto, M., Tesfai, A., Fields, A., Kline, K., Fiddner, J., Higa, J., Barnes, A., Arroyo, F., Salvatierra, A., Holland, A., … Gieraltowski, L. (2019). Shiga toxin–producing *Escherichia coli* infections associated with romaine lettuce—United States, 2018. *Clinical Infectious Diseases,* April 2018. https://doi.org/10.1093/cid/ciz1182

Buchholz, U., Bernard, H., Werber, D., Böhmer, M. M., Remschmidt, C., Wilking, H., Deleré, Y., An Der Heiden, M., Adlhoch, C., Dreesman, J., Ehlers, J., Ethelberg, S., Faber, M., Frank, C., Fricke, G., Greiner, M., Höhle, M., Ivarsson, S., Jark, U., … Kühne, M. (2011). German outbreak of *Escherichia coli* O104:H4 associated with sprouts. *New England Journal of Medicine.* https://doi.org/10.1056/NEJMoa1106482

Burger, R. (2012). EHEC O104:H4 in Germany 2011: large outbreak of bloody diarrhea and haemolytic uraemic syndrome by Shiga toxin–producing *E. coli* via contaminated food. In E. R. Choffnes, D. A. Relman, L. Olsen, R. Hutton, & A. Mack (Eds.), *Improving Food Safety through a One Health Approach.* National Academies Press. https://doi.org/10.17226/13423

Calisher, C., Carroll, D., Colwell, R., Corley, R. B., Daszak, P., Drosten, C., Enjuanes, L., Farrar, J., Field, H., Golding, J., Gorbalenya, A., Haagmans, B., Hughes, J. M., Karesh, W. B., Keusch, G. T., Lam, S. K., Lubroth, J., Mackenzie, J. S., Madoff, L., … Turner, M. (2020). Statement in support of the scientists, public health professionals, and medical professionals of China combatting COVID-19. *Lancet.* https://doi.org/10.1016/S0140-6736(20)30418-9

Canadian Food Inspection Agency. (2019). *Food Safety Investigation and Recall Process.* https://www.inspection.gc.ca/food-safety-for-industry/information-for-consumers/food-safety-system/basic-html/eng/1374439778888/1374821384212

Centers for Disease Control and Prevention. (2016a). *Outbreak Detection.* https://www.cdc.gov/pulsenet/outbreak-detection/index.html

Centers for Disease Control and Prevention. (2016b). *PulseNet.* https://www.cdc.gov/pulsenet/index.html

Centers for Disease Control and Prevention. (2018a). *Estimates of Foodborne Illness in the United States.* https://www.cdc.gov/foodborneburden/index.html

Centers for Disease Control and Prevention. (2018b). *Foodborne Disease Outbreak Surveillance System (FDOSS)*. https://www.cdc.gov/fdoss/index.html

Centers for Disease Control and Prevention. (2018c). *Foodborne Outbreak Response Team*. https://www.cdc.gov/ncezid/dfwed/orpb/ort.html

Centers for Disease Control and Prevention. (2018d). *Multistate Outbreak of E. coli O157:H7 Infections Linked to Romaine Lettuce (Final Update)*. https://www.cdc.gov/ecoli/2018/o157h7-04-18/index.html

Centers for Disease Control and Prevention. (2018e). *Steps in a Foodborne Outbreak Investigation*. https://www.cdc.gov/foodsafety/outbreaks/investigating-outbreaks/investigations/index.html

Centers for Disease Control and Prevention. (2019a). *About PulseNet*. https://www.cdc.gov/pulsenet/about/index.html

Centers for Disease Control and Prevention. (2019b). *About the Environmental Health Specialists Network (EHS-Net)*. https://www.cdc.gov/nceh/ehs/ehsnet/about.htm

Centers for Disease Control and Prevention. (2019c). *Foodborne Diseases Active Surveillance Network (FoodNet)*. https://www.cdc.gov/foodnet/index.html

Centers for Disease Control and Prevention. (2019d). *Foodborne Diseases Centers for Outbreak Response Enhancement*. https://www.cdc.gov/foodcore/index.html

Centers for Disease Control and Prevention. (2019e). *Integrated Food Safety Centers of Excellence*. https://www.cdc.gov/foodsafety/centers/

Centers for Disease Control and Prevention. (2019f). *National Notifiable Diseases Surveillance System (NNDSS)*. https://wwwn.cdc.gov/nndss/

Centers for Disease Control and Prevention. (2020). *National Surveillance of Bacterial Foodborne Illnesses (Enteric Diseases)*. https://www.cdc.gov/nationalsurveillance/index.html

Chan, J. F. W., Yuan, S., Kok, K. H., To, K. K. W., Chu, H., Yang, J., Xing, F., Liu, J., Yip, C. C. Y., Poon, R. W. S., Tsoi, H. W., Lo, S. K. F., Chan, K. H., Poon, V. K. M., Chan, W. M., Ip, J. D., Cai, J. P., Cheng, V. C. C., Chen, H., … Yuen, K. Y. (2020). A familial cluster of pneumonia associated with the 2019 novel coronavirus indicating person-to-person transmission: a study of a family cluster. *Lancet*. https://doi.org/10.1016/S0140-6736(20)30154-9

Council to Improve Foodborne Outbreak Response. (2014). *Guidelines for Foodborne Disease Outbreak Response, Second Edition*.

Croxen, M. A., Law, R. J., Scholz, R., Keeney, K. M., Wlodarska, M., & Finlay, B. B. (2013). Recent advances in understanding enteric pathogenic *Escherichia coli*. *Clinical Microbiology Reviews*, 26(4), 822–880. https://doi.org/10.1128/CMR.00022-13

European Centre for Disease Prevention and Control. (2020a). *European Food- and Waterborne Diseases and Zoonoses Network (FWD-Net)*. https://www.ecdc.europa.eu/en/about-us/partnerships-and-networks/disease-and-laboratory-networks/fwd-net

European Centre for Disease Prevention and Control. (2020b). *Food- and Waterborne Diseases and Zoonoses Programme*. https://www.ecdc.europa.eu/en/about-uswho-we-aredisease-programmes/food-and-waterborne-diseases-and-zoonoses-programme

European Food Safety Authority and European Centre for Disease Prevention and Control. (2019). The European Union One Health 2018 Zoonoses Report. *EFSA Journal*, 17(12). https://doi.org/10.2903/j.efsa.2019.5926

European Food Safety Authority. (2011). Tracing seeds, in particular fenugreek (*Trigonella foenum-graecum*) seeds, in relation to the Shiga toxin-producing *E. coli* (STEC) O104:H4 2011 outbreaks in Germany and France. *EFSA Supporting Publications, 8*(7). https://doi.org/10.2903/sp.efsa.2011.EN-176

Gault, G., Weill, F. X., Mariani-Kurkdjian, P., Jourdan-da Silva, N., King, L., Aldabe, B., Charron, M., Ong, N., Castor, C., Macé, M., Bingen, E., Noël, H., Vaillant, V., Bone, A., Vendrely, B., Delmas, Y., Combe, C., Bercion, R., D'Andigné, E., … Rolland, P. (2011). Outbreak of haemolytic uraemic syndrome and bloody diarrhoea due to *Escherichia coli* O104:H4, south-west France, June 2011. *Eurosurveillance*. https://doi.org/10.2807/ese.16.26.19905-en

Gossner, C. M. (2016). New version of the epidemic Intelligence Information System for food- and waterborne diseases and zoonoses (EPIS-FWD) launched. *Eurosurveillance, 21*(49), 30422. https://doi.org/10.2807/1560-7917. ES.2016.21.49.30422

Henao, O. L., Jones, T. F., Vugia, D. J., & Griffin, P. M. (2015). Foodborne diseases active surveillance network—2 decades of achievements, 1996–2015. *Emerging Infectious Diseases*. https://doi.org/10.3201/eid2109.150581

Hunter-Adams, J., Battersby, J., & Oni, T. (2018). Fault lines in food system governance exposed: reflections from the *Listeria* outbreak in South Africa. *Cities & Health*. https://doi.org/10.1080/23748834.2018.1508326

Kearney, G. D. (2018). Introduction to foodborne illness outbreak investigations. In *Environmental Public Health: The Practitioner's Guide*. American Public Health Association. https://doi.org/10.2105/9780875532943ch13

Knödler, M., Berger, M., & Dobrindt, U. (2016). Long-term survival of the Shiga Toxin-producing *Escherichia coli* O104: H4 outbreak strain on fenugreek seeds. *Food Microbiology*. https://doi.org/10.1016/j.fm.2016.06.005

Marvin, H. J. P., Kleter, G. A., Prandini, A., Dekkers, S., & Bolton, D. J. (2009). Early identification systems for emerging foodborne hazards. *Food and Chemical Toxicology, 47*(5), 915–926. https://doi.org/10.1016/j.fct.2007.12.021

Moorthy, V., Henao Restrepo, A. M., Preziosi, M.-P., & Swaminathan, S. (2020). Data sharing for novel coronavirus (COVID-19). *Bulletin of the World Health Organization, 98*(3), 150–150. https://doi.org/10.2471/BLT.20.251561

Nadon, C., Van Walle, I., Gerner-Smidt, P., Campos, J., Chinen, I., Concepcion-Acevedo, J., Gilpin, B., Smith, A. M., Kam, K. M., Perez, E., Trees, E., Kubota, K., Takkinen, J., Nielsen, E. M., & Carleton, H. (2017). PulseNet International: Vision for the implementation of whole genome sequencing (WGS) for global food-borne disease surveillance. *Eurosurveillance, 22*(23), 30544. https://doi. org/10.2807/1560-7917.ES.2017.22.23.30544

Public Health Agency of Canada. (2013a). *Centre for Food-borne, Environmental and Zoonotic Infectious Diseases — Canada.c.* https://www.canada.ca/en/public-health/services/infectious-diseases/centre-food-borne-environmental-zoonotic-infectious-diseases.html

Public Health Agency of Canada. (2013b). *Overview of FoodNet Canada: Reducing the Burden of Gastrointestinal Disease in Canada — Canada.ca.* https://www. canada.ca/en/public-health/services/surveillance/foodnet-canada/overview. html

Public Health Agency of Canada. (2017). *Canada's Foodborne Illness Outbreak Response Protocol (FIORP) — A Guide to Multi-Jurisdictional Enteric Outbreak Response.*

Public Health Agency of Canada. (2018). *National Enteric Surveillance Program (NESP).* https://www.canada.ca/en/public-health/programs/national-enteric-surveillance-program.html

Public Health Agency of Canada. (2019a). *National Microbiology Laboratory.* https://www.canada.ca/en/public-health/programs/national-microbiology-laboratory.html

Public Health Agency of Canada. (2019b). *PulseNet Canada.* https://www.canada.ca/en/public-health/programs/pulsenet-canada.html

Public Health Agency of Canada. (2020). *Notifiable Diseases Online.* https://dsol-smed.phac-aspc.gc.ca/notifiable/

Ramaswamy, V., Cresence, V. M., Rejitha, J. S., Lekshmi, M. U., Dharsana, K. S., Prasad, S. P., & Vijila, H. M. (2007). *Listeria* — review of epidemiology and pathogenesis. *Journal of Microbiology, Immunology and Infection, 40*(1), 4–13.

Rohde, H., Qin, J., Cui, Y., Li, D., Loman, N. J., Hentschke, M., Chen, W., Pu, F., Peng, Y., Li, J., Xi, F., Li, S., Li, Y., Zhang, Z., Yang, X., Zhao, M., Wang, P., Guan, Y., Cen, Z., … Yang, R. (2011). Open-source genomic analysis of Shiga-toxin–producing. *New England Journal of Medicine.*

Romaine Task Force. (2019). *Final Report and Recommendations.* https://www.unitedfresh.org/content/uploads/2019/09/Final-Romaine-Task-Force-Report-9-30-18.pdf

Scallan, E., Hoekstra, R. M., Angulo, F. J., Tauxe, R. V., Widdowson, M. A., Roy, S. L., Jones, J. L., & Griffin, P. M. (2011). Foodborne illness acquired in the United States — Major pathogens. *Emerging Infectious Diseases.* https://doi.org/10.3201/eid1701.P11101

Seitzinger, P. J., Tataryn, J., Osgood, N., & Waldner, C. (2019). Foodborne outbreak investigation: effect of recall inaccuracies on food histories. *Journal of Food Protection, 82*(6), 931–939. https://doi.org/10.4315/0362-028X.JFP-18-548

Smith, A. M., Tau, N. P., Smouse, S. L., Allam, M., Ismail, A., Ramalwa, N. R., Disenyeng, B., Ngomane, M., & Thomas, J. (2019). Outbreak of *Listeria* Monocytogenes in South Africa, 2017 – 2018: laboratory activities and experiences associated with whole-genome sequencing analysis of isolates. *Foodborne Pathogens and Disease.* https://doi.org/10.1089/fpd.2018.2586

Thomas, J., Govender, N., McCarthy, K. M., Erasmus, L. K., Doyle, T. J., Allam, M., Ismail, A., Ramalwa, N., Sekwadi, P., Ntshoe, G., Shonhiwa, A., Essel, V., Tau, N., Smouse, S., Ngomane, H. M., Disenyeng, B., Page, N. A., Govender, N. P., Duse, A. G., … Blumberg, L. H. (2020). Outbreak of Listeriosis in South Africa associated with processed meat. *New England Journal of Medicine.* https://doi.org/10.1056/NEJMoa1907462

Todd, E. C. D. (2011). Procedures to investigate foodborne illness. In *Procedures to Investigate Foodborne Illness* (pp. 1–164). Springer US. https://doi.org/10.1007/978-1-4419-8396-1_1

U.S. Food and Drug Administration. (2018). *Memorandum to the File on the Environmental Assessment; Yuma 2018 E. coli O157:H7 Outbreak Associated with Romaine Lettuce.* https://www.fda.gov/media/120690/download

U.S. Food and Drug Administration. (2019). *Rapid Response Teams (RRTs).* https://www.fda.gov/federal-state-local-tribal-and-territorial-officials/national-integrated-food-safety-system-ifss-programs-and-initiatives/rapid-response-teams-rrts

U.S. Food and Drug Administration. (2020a). *About the CORE Network.* https://www.fda.gov/food/outbreaks-foodborne-illness/about-core-network

U.S. Food and Drug Administration. (2020b). *Coronavirus Disease 2019 (COVID 19): Frequently Asked Questions.* https://www.fda.gov/emergency-preparedness-and-response/mcm-issues/coronavirus-disease-2019-covid-19-frequently-asked-questions#food

United Federal State & Local Laboratories for Food Emergency Response. (2019). *Food Emergency Response Network (FERN)*. https://www.fernlab.org/

United States Department of Agriculture. (2019). *About FSIS*. https://www.fsis.usda.gov/wps/portal/informational/aboutfsis

Werber, D., Krause, G., Frank, C., Fruth, A., Flieger, A., Mielke, M., Schaade, L., & Stark, K. (2012). Outbreaks of virulent diarrheagenic *Escherichia coli* - are we in control? *BMC Medicine*. https://doi.org/10.1186/1741-7015-10-11

Williams, I. T., Whitlock, L., & Wise, M. E. (2019). Acute enteric disease outbreaks. In S. A. Rasmussen & R. A. Goodman (Eds.), *The CDC Field Epidemiology Manual* (Issue March 2020). Oxford University Press. https://doi.org/10.1093/oso/9780190933692.001.0001

World Health Organization. (2008). *Foodborne Disease Outbreaks: Guidelines for Investigation and Control*.

World Health Organization. (2015). *WHO Estimates of the Global Burden of Foodborne Diseases*.

World Health Organization. (2019). *More Complex Foodborne Disease Outbreaks Require New Technologies, Greater Transparency*. https://www.who.int/news-room/detail/06-12-2019-more-complex-foodborne-disease-outbreaks-requires-new-technologies-greater-transparency

World Health Organization. (2020a). *Coronavirus Disease 2019 (COVID-19) Situation Report–52*. https://www.who.int/docs/default-source/coronaviruse/20200312-sitrep-52-covid-19.pdf?sfvrsn=e2bfc9c0_2

World Health Organization. (2020b). *Global Foodborne Infections Network (GFN)*. https://www.who.int/foodsafety/areas_work/foodborne-diseases/gfn/en/

chapter six

Impact of the SARS-CoV-2 (COVID-19) pandemic on the United States food system

Adam Friedlander, Lily Yang, Nicole Arnold, and Stephanie Brown

Contents

EDITORS' NOTE: EVOLVING SCIENCE DURING THE COVID-19 PANDEMIC

This chapter represents a brief overview of events, current knowledge, and best practices associated with SARS-CoV-2, the virus which causes COVID-19. At this time (October 2020), there is no approved vaccine or therapeutic for SARS-CoV-2 in the United States. The science and understanding of SARS-CoV-2 is continuing to evolve as more is learned through studying this pathogen and its characteristics. Therefore, information presented in this chapter only reflects current status or understanding of SARS-CoV-2, and may not reflect future information. The perspectives and opinions within this chapter are those of the authors and are not representative of the views of affiliated organizations.

Introduction of SARS-CoV-2

Infection with severe acute respiratory syndrome coronavirus 2 (SARS-CoV-2), a highly contagious respiratory virus, causes the novel 2019 coronavirus disease (COVID-19) (WHO, 2020a,b). As of October 2020, there have been over 37 million reported cases of COVID-19 and over 1 million deaths, globally (Johns Hopkins University and Medicine, 2020). People living in the United States represent nearly 20 percent of all global COVID-19 illnesses and deaths. To date, over 6 million people living in the United States have tested positive for COVID-19, an unknown number have been hospitalized, and over 200,000 people have died (Johns Hopkins University and Medicine, 2020).

People infected with SARS-CoV-2 have reported a wide range of symptoms, ranging from mild to severe (U.S. Centers for Disease Control and Prevention, 2020a). The Centers for Disease Control and Prevention (CDC) names at least 11 symptoms used to determine if someone may have COVID-19, including (U.S. Centers for Disease Control and Prevention, 2020a):

- Fever or chills
- Cough
- Shortness of breath or difficulty breathing
- Fatigue
- Muscle or body aches
- Headache
- New loss of taste or smell
- Sore throat
- Congestion or runny nose
- Nausea or vomiting
- Diarrhea

These symptoms are shared with other illnesses, including the common cold, influenza, foodborne illness, and a variety of other illnesses. Therefore, it may be difficult to determine if someone has COVID-19 without confirmation by viral testing. People with certain underlying medical conditions, including susceptible populations, have an increased risk for severe illness if they contract SARS-CoV-2 (U.S. Centers for Disease Control and Prevention, 2020b). There are also identified underlying conditions and behaviors that might result in an increased risk of illness from SARS-CoV-2 (e.g., asthma, pregnancy, and smoking) (U.S. Centers for Disease Control and Prevention, 2020b). Conditions such as cancer, chronic kidney disease, obesity, and COPD (chronic obstructive pulmonary disease) also increase the risk of severe illness from SARS-CoV-2 (U.S. Centers for Disease Control and Prevention, 2020b).

Infected individuals may also be asymptomatic, meaning that they do not display any symptoms. Preliminary research estimates that up to 40 percent of COVID-19 cases are asymptomatic (U.S. Centers for Disease Control and Prevention, 2020c). Asymptomatic individuals may be unaware that they are carriers of the disease, subsequently increasing the spread of illness if preventive measures are not followed.

The broadscale economic and social outcomes of the pandemic are currently unknown. A combined study between City University of New York (CUNY) Graduate School of Public Health and Health Policy, the Infectious Disease Clinical Outcomes Research Unit at the Los Angeles Biomedical Research Institute, Harbor-UCLA Medical Center, and Torrance Memorial Medical Center estimated in April 2020 that symptomatic COVID-19 individuals can have up to $3,000 in direct medical costs, which is four times higher than the medical costs of symptomatic influenza cases (CUNY Graduate School of Public Health and Health Policy, 2020). Some individuals who contract SARS-CoV-2 have persistent symptoms or health effects that can create additional economic and health burdens for the individual, their community, and workplace (U.S. Centers for Disease Control and Prevention, 2020d). There is increasing evidence that the short- and long-term health effects (e.g., myocarditis, heart damage) associated with COVID-19 can delay full recovery for sickened individuals (Arons et al., 2020; U.S. Centers for Disease Control and Prevention, 2020e). Additionally, systemic health disparities/inequities, defined as "a particular type of difference in health" in which "disadvantaged social groups […] systematically experience worse health or greater health risks than more advantaged social groups," became increasingly evident as the pandemic persisted (Braveman, 2006; Price et al., 2020; Khazanchi et al., 2020; Goyal et al., 2020; Kim et al., 2020; U.S. Centers for Disease Control and Prevention, 2020f). These disparities included structural racism, discrimination, lack of educational opportunities and material access, and inaccessibility to medical aid, which further contributed to the risks of complications from COVID-19 (Braveman, 2006; Baciu et al., 2017; U.S. Centers for Disease Control and Prevention, 2020g,h).

Impacts of SARS-CoV-2

In an effort to flatten the curve and limit the spread of SARS-CoV-2, preventive measures were advised by public health officials and implemented at global, national, state, and local levels. Preventive measures included movement restrictions (e.g., lockdowns, stay-at-home-orders, temporary business closings, and reduced business hours of operation), mask-wearing recommendations and/or mandates, limitations on human gatherings, social distancing (maintaining a minimum of 6-feet/2-meter distance) measures, promotion of good personal hand hygiene (washing hands frequently, using 60 percent-alcohol-based hand sanitizers when handwashing is unavailable), respiratory hygiene (covering one's mouth when coughing, sneezing into elbows), and frequent cleaning and disinfecting procedures for high-touch surfaces.

While many businesses shut down as the pandemic spread, starting in March 2020, food production was still necessary to feed the population. To continue to supply a locked-down nation with food, the U.S. food and agriculture sectors were among the primary industries deemed, by the U.S. Department of Homeland Security (DHS) and the White House, as "critical infrastructure." Critical infrastructures are any "systems and assets, whether physical or virtual, so vital to the United States that the incapacity or destruction of such systems and assets would have a debilitating impact on security, national economic security, national public health or safety, or any combination of those matters" (U.S. Cybersecurity and Infrastructure Security Agency, 2020). To safely feed the nation, food and agriculture workers directly involved with food production and supply, as part of the nation's "critical infrastructure," were considered "essential workers" at state, local, and tribal levels. These food and agriculture workers comprised approximately 21 percent of the overall "essential" workforce (U.S. Cybersecurity and Infrastructure Security Agency, 2020; EPI, 2020). Within the food and agriculture sectors, essential workers included a wide range of personnel, including growers, farm workers, foodservice workers, food facility employees, distributors, truck drivers, sanitation workers, grocery store clerks, and other participants within the food supply chain. While workers in other industries (e.g., service, travel, and hospitality) were mandated by many local and state ordinances from physically attending work at the beginning of the pandemic, a designated segment of food and agricultural workers continued to work on-site to protect national interests.

With the onset of the pandemic, the U.S. food system was heavily impacted through market disruptions, product shortages (e.g., cleaning and disinfecting supplies, paper products, hand sanitizers, and food items and ingredients), inabilities to redirect food distribution and supply chains, and workforce disruptions due to food worker health (e.g., illness, death, or employee exclusion due to potential contact with an infected individual). As essential workers ensured the continuation of the food supply, they were faced with obvious health risks associated with contracting SARS-CoV-2; over 1,000 essential workers across all critical U.S. industries have died from COVID-19 (New York State, 2020; Amnesty International, 2020).

Additionally, the racial disparities brought about by the pandemic were accentuated within the food and agriculture sector. Minority populations and Persons of Color make up approximately 50 percent of all food and agriculture essential workers, while making less per hour than their

White counterparts (EPI, 2020; United States Department of Agriculture Economic Research Service, 2020). Moreover, up to 60 percent of farm laborers are from minority groups and about half of all hired crop farm workers lack legal U.S. immigration status (EPI, 2020; United States Department of Agriculture Economic Research Service, 2020). Studies from 2020 have found that Black, Indigenous, Persons of Color, and Hispanic individuals have and continue to experience an excess burden of SARS-CoV-2 infection (Price et al., 2020; Khazanchi et al., 2020; Goyal et al., 2020; Kim et al., 2020; U.S. Centers for Disease Control and Prevention, 2020f; Rentsch et al., 2020; UN, 2020a). Vulnerable populations (as it pertains to food and agriculture workers) include individuals in temporary employment positions, individuals without legal immigration status, and those that are unable to "shelter at home" who must be present at the workplace (United States Department of Agriculture Economic Research Service, 2020; Center for American Progress, 2020). Vulnerable food and agriculture individuals may also accrue low pay, have lower rates of health insurance coverage, live in multigenerational accommodations (thereby increasing the risk of COVID-19 exposure to elderly or health-averse individuals), or temporary housing situations (European Commission, 2020; Center for American Progress, 2020). These factors increase their risks of contracting and spreading this communicable disease.

At the onset of the COVID-19 pandemic, characteristics and practices inherent in food and agricultural workplaces were attributed to further spread of COVID-19. For example, prior to the COVID-19 pandemic, those working in food processing operations often found themselves in close-contact with other employees for prolonged periods of time, farm workers were in close contact during harvest and also in employee-provided accommodations, and essential retail and food service workers found themselves in direct contact with customers who were initially unaware of the infectiousness of SARS-CoV-2 (North Dakota Department of Health, 2020).

These on-site essential workers were confronted with increasing challenges to protect themselves, their families, and their income in the face of increasing public health dangers (*Lancet*, 2020). However, working while ill, although discouraged, continues to occur as one's livelihood is typically dependent on wages. Due to the food and agriculture industry's fast-paced work process, employee policies, and the need to support one's livelihood, sick leave use by asymptomatic or symptomatic individuals may not be utilized or provided. Currently, if an individual is considered an essential worker, it is recommended that a worker who has contacted SARS-CoV-2 be isolated for at least 10 days prior to returning to work (Food and Beverage Issue Alliance, 2020; U.S. Centers for Disease Control and Prevention, 2020i). However, if a worker is paid on an hourly basis, not working for 10-or-more days can equate to a significant loss in income.

To address the shortage of essential workers, the CDC has further recommended that essential workers, if critical for a business, may return to
work with one or two negative SARS-CoV-2 tests; however, this practice,
while acceptable, may put other workers at risk (U.S. Centers for Disease
Control and Prevention, 2020i).

At the beginning of the pandemic, on-site essential workers found
themselves without adequate personal protective equipment (PPE) to
protect themselves at work based on national shortages. The following is
Food and Drug Administration's (FDA) summation on the PPE shortages
that occurred in the early months of the COVID-19 pandemic (U.S. Food
and Drug Administration, 2020a).

> The COVID-19 pandemic has caused shortages in
> PPE, cloth face coverings, disinfectants, and sanita-
> tion supplies. Some food production stakeholders
> face challenges procuring adequate materials due
> to increased demand across all economic sectors
> and the disruption of traditional supply chains.
> These challenges may, in turn, cause interruptions
> in the food supply chain. Disruptions to the supply
> of food and agricultural commodities or the inputs
> needed to produce them, however minor, can cause
> significant ripple effects through all nodes in the
> supply chain, from growers and suppliers to pro-
> cessors and distributors to retailers.

As high-profile situations in the food industry (e.g., widespread viral
dissemination in meat processing plants) arose at the beginning of the
pandemic, essential food and agriculture workers received heightened
recognition during the COVID-19 pandemic from government officials,
public health experts, industry professionals, and consumers. Resulting
from the increased publicity and the public's awareness of the food system, the food industry was compelled to implement more safeguards
to protect essential food and agricultural workers' health through preventive measures, with the ultimate goal of protecting public health.
Examples of these preventive measures included staggering work shifts,
limiting the number of individuals allowed inside a facility, mandating
mask use in the workplace, enforcing social distancing within workspaces
to protect employees and customers, increased handwashing stations and
hand sanitizer supplies, providing engineering controls such as plexiglass
in situations where social distancing was unfeasible, encouragement of
working-from-home when feasible, and providing more sick leave (U.S.
Centers for Disease Control and Prevention, 2015). These preventive measures, specifically social distancing, mask wearing, hand washing, and

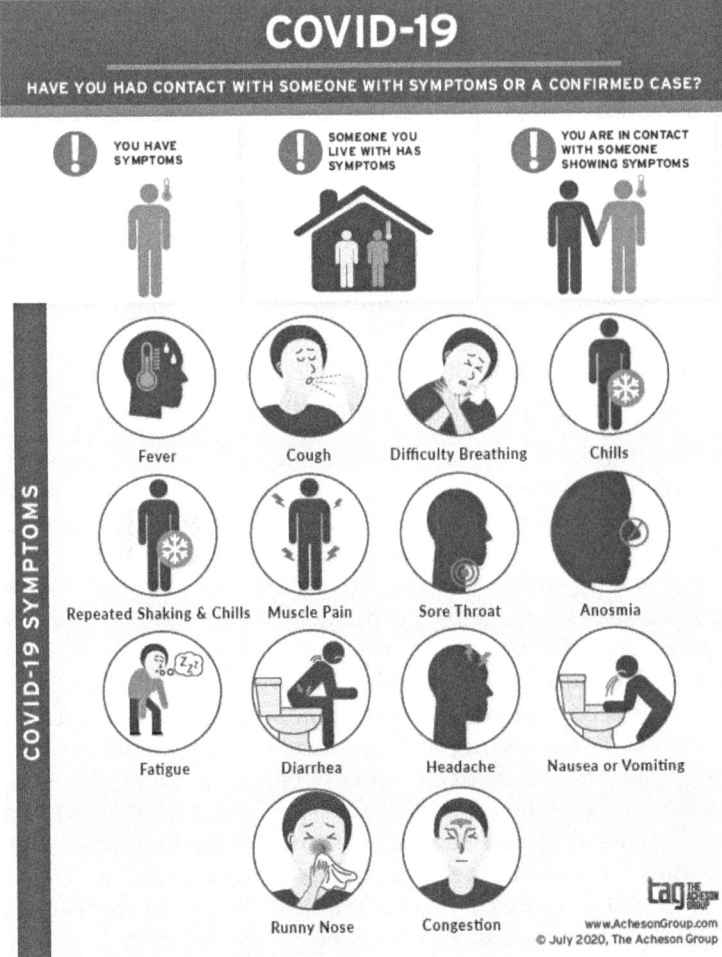

increased cleaning and sanitation, became the "standards" to protect both workers and the public, alike. Ultimately, the health and safety of those working in critical infrastructure must be considered, especially essential workers who are part of vulnerable work and/or susceptible populations.

SARS-CoV-2 and protecting workers in a food setting

Although SARS-CoV-2 is an infectious, communicable disease, the U.S. CDC, European Centre for Disease Prevention and Control, International Commission on Microbiological Specifications for Foods (ICMSF), and the

World Health Organization (WHO) all agree that there is no evidence to suggest that handling or consuming food is associated with COVID-19 (U.S. Centers for Disease Control and Prevention, 2020j; ICMSF, 2020). Yet, because SARS-CoV-2 can be transmitted through other routes in a food production environment, many of the controls used to reduce the risk of foodborne illnesses (e.g., personal hygiene, frequent handwashing, enhanced cleaning and sanitation, exclusion of ill employees) can also be used to control the spread of SARS-CoV-2, including on fomites and/or through droplets and aerosols in the environment (WHO, 2020c; Meselson, 2020).

Cleaning and sanitation as a preventive measure for food safety and SARS-CoV-2 mitigation

During the pandemic, implementation of enhanced cleaning and sanitation programs varied depending on business entity and regulatory requirements. Food operations personnel, trade associations, academic groups, consultants, and governmental organizations in the farm-to-fork continuum (e.g., growers, manufacturers, distributors, markets, grocery stores, and last-mile food delivery) promoted health information and protocols to equip individuals with preventive measures to protect against COVID-19. This information included, but was not limited to: (1) encouraging workers and customers to stay home when sick via self-isolation or quarantine; (2) handwashing with soap and water for at least 20 seconds; (3) using a 60 percent or greater alcohol-based sanitizer (depending on alcohol-base used) when handwashing was not available; (4) performing routine cleaning and sanitation activities for high-touch surfaces; (5) updating air ventilation systems; (6) wearing a cloth face covering at work and in public spaces; (7) social/physical distancing at 6 feet (or more) from others that do not live in one's home; (8) not touching one's face; (9) participating in contact tracing investigations; and (10) receiving a viral test if suspected of illness. Yet, it is unknown whether these measures, in turn, encouraged and promoted essential food workers to extend these precautionary measures in their own spaces.

To protect consumer health and to comply with regulatory requirements, members of the U.S. food system have food safety systems in place that contain cleaning and sanitation practices and programs. Well-designed and properly implemented cleaning and sanitation programs can reduce the risk of foodborne illness by preventing and/or reducing contamination events from occurring in food production environments. With the emergence of the COVID-19 pandemic, the food and agriculture industry added steps to their cleaning and sanitation or other workplace programs to reduce viral particle transmission between workers in a facility. These additional steps were designed to protect the current workforce from respiratory illness, unlike other cleaning and sanitation measures, which are designed to reduce risks of consumer illness with foodborne pathogens (e.g., *Listeria monocytogenes, Escherichia coli*, and *Salmonella*).

Effective cleaning and sanitation practices are valuable both in production environments and in community and home settings. Employees such as teachers/professors, daycare/childcare employees, and those residing in office spaces took on cleaning and sanitation roles and responsibilities outside of their typical employee responsibilities in the COVID-19 era. For all settings, understanding and implementing proper cleaning and sanitizing procedures on surfaces will mitigate pathogens of concern (e.g., foodborne pathogens, SARS-CoV-2). Effective cleaning and sanitation in a wet environment have four basic steps including a prerinse, cleaning with a detergent, rinse, and application of a sanitizer (Stone et al., 2020). The first three steps would fall within the cleaning portion of a cleaning and sanitation program. Cleaning, as defined by the CDC, involves the "physical removal of foreign material (e.g., dust, soil) and organic material (e.g., blood, secretions, excretions, microorganisms). Cleaning physically removes, rather than kills, most microorganisms of concern from a surface. It is accomplished with water, detergents, and mechanical action" (U.S. Centers for Disease Control and Prevention, 2020k). On the other hand, sanitation programs (which follow cleaning programs) include using sanitizers or disinfectants to reduce and/or inactivate remaining microorganisms after cleaning steps. As defined by the Environmental Protection Agency (EPA) under 40 CFR §158.2203, "sanitizer" means "a substance, or mixture of substances, that reduces the bacteria population in the inanimate environment by significant numbers but does not destroy or eliminate all bacteria. Sanitizers meeting Public Health Ordinances are generally used on food contact surfaces and are termed sanitizing rinses" (40 CFR §158.2203). Under the same statute, "disinfectants" are defined as a "substance, or mixture of substances, that destroys or irreversibly inactivates bacteria, fungi and viruses, but not necessarily bacterial spores, in the inanimate environment" (40 CFR §158.2203). In the United States, products labeled as cleaning agents (i.e., detergents that make a pesticidal claim), sanitizers, and disinfectants cannot be used unless specifically registered with the EPA as antimicrobial pesticides (U.S. Environmental Protection Agency, 2020a,b).

The EPA, an agency within the Executive Branch, reviews, registers, and regulates chemical "pesticides" effective against pathogens of concern, including viral respiratory pathogens such as SARS-CoV-2 and foodborne pathogens such as norovirus (U.S. Environmental Protection Agency, 2020c). EPA's authority to regulate antimicrobial pesticides such as sanitizers and disinfectants (i.e., common types of pesticides) derives from the Federal Insecticide, Fungicide, and Rodenticide Act (FIFRA) of 1996 (7 U.S.C. §136 et seq), explained as (U.S. Environmental Protection Agency, 2020b):

> [A] "pesticide" [is defined] as "any substance or mixture of substances intended for preventing, destroying, repelling, or mitigating any pest." FIFRA § 2(u), 7 U.S.C. § 136(u). Unless otherwise exempted from registration under 40 CFR §§ 152.20, 152.25 or 152.30, pesticide products that are intended for a pesticidal purpose must be registered. A product is considered to be intended for a pesticidal purpose if, among other things, the person who distributes or sells it claims, states, or implies that the product prevents, destroys, repels or mitigates a pest. Therefore, once a product label (or other statement made in connection with the sale or distribution of the product) includes any claim of pest mitigation, under 40 CFR § 152.15, the product is one that is intended for a pesticidal purpose and becomes subject to the registration provisions of FIFRA.

While cleaning and sanitation programs can reduce the risk of spreading pathogens, they are most effective when chemical products are used properly by the person performing cleaning and sanitation tasks. Proper usage includes, but is not limited to (Stone et al., 2020):

- Ensuring that the product can be used in one's region. Even if a product is registered for use nationally, local and state regulations may be more restrictive with certain pesticide applications.
- Precisely following instructions on the label.
- Not mixing products unless stated on the label.
- Proper labeling if chemicals are transferred to a secondary container.
- Wearing personal protective equipment (e.g., gloves, goggles, mask) as directed by product label.
- Cleaning surfaces prior to the application of any sanitizer or disinfectant products.
- Applying products on approved surfaces for recommended contact time, as directed by product label.
- Ensuring proper ventilation, as directed by provided label.
- Safely storing products so they are inaccessible to nonapproved users.
- Handwashing after handling any antimicrobial pesticides.

Labels of products used

One of the most important aspects of proper cleaning and sanitation, regardless of the user, is understanding the product's label. To improve sanitation protocols in critical businesses and communities throughout the COVID-19 pandemic, the need to select and use disinfectants effective against SARS-CoV-2 became a necessary component to combat this pathogen. As a result of heightened awareness and pressure to use pesticides, as defined under 40 CFR §158.2203, the EPA developed a SARS-CoV-2-specific database, titled *List N: Products with Emerging Viral Pathogens AND Human Coronavirus claims for use against SARS-CoV-2*, which can be found online and utilized by anyone (U.S. Environmental Protection Agency, 2020d). To be added to List N, disinfectants were evaluated by the EPA to successfully (U.S. Environmental Protection Agency, 2020e):

1. demonstrate efficacy against the coronavirus SARS-CoV-2;
2. demonstrate efficacy against a pathogen that is harder to kill than SARS-CoV-2; or
3. demonstrate efficacy against a different human coronavirus similar to SARS-CoV-2.

To find a specific SARS-CoV-2-effective product on the EPA List N, an individual must first identify the EPA registration number on the label. If a product does not contain an EPA registration number and makes a pesticide claim, the product is not able to be used under Section 12 of FIFRA (U.S. Environmental Protection Agency, 2020f). Once the EPA number has been identified, the first two parts of the registration number are used to match the primary registration number on the EPA's public database. If a match is found, then the product is eligible for use against SARS-CoV-2 on specific surfaces as stated by the product's label. For example, "[If] EPA Reg. No. 12345-12 is on List N, you can buy EPA Reg. No. 12345-12-2567 and know you're getting an equivalent product" (U.S. Environmental Protection Agency, 2020g,h). However, even if a product is not available on List N, but its registration number indicates that human coronavirus is a targeted pathogen, then the EPA considers the product appropriate for use against SARS-CoV-2 (U.S. Environmental Protection Agency, 2020i).

Since commercial disinfectants are delivered in unique product formulations, the EPA must approve the ingredients and formulations based on its specific purpose and how the product will be used. The List N database allows individuals to search for SARS-CoV-2 specific disinfectants based upon its formulation type. Some examples of formulation types for specific disinfectant delivery mechanisms include dilutable, electrostatic spray, fog, mist, ready-to-use, wipe. However, when disinfectants effective against SARS-CoV-2 are approved, this does not equate to all delivery mechanisms being approved as well. Therefore, it is imperative to look at List N when considering which disinfectants and application method(s) will be used to reduce SARS-CoV-2 in your food processing environment. Failure to follow the manufacturer's label instructions on EPA-registered pesticides can lead to harmful health outcomes, such as mild or severe chemical poisoning or damage to equipment (e.g., fomite discolorations and corrosion, payment terminal damages).

Ultimately, the onus of identifying the correct products to use falls upon the end-user. Both in the food/agriculture industry and for consumers, this could be difficult if the end-user does not have the resources or technical background to understand product labels. To compound the difficulty, end-users may not be aware of the importance of safe chemical handling practices or aware of EPA List N, which is specifically for SARS-CoV-2 products. Additionally, even if a product contained an active and approved ingredient that was found in other products on the EPA List N and effective against SARS-CoV-2, the product itself may not be approved by the EPA for use against SARS-CoV-2, leading to even further end-user confusion. These issues, which were exacerbated during the COVID-19 pandemic, show the importance of cleaning and sanitation education for all end-users.

Documentation of pesticide usage

Food and agriculture workers who execute cleaning and sanitation procedures in a food operation must be trained to ensure proper procedures are followed and workers are protected during usage. In the food and agriculture industry, properly documented cleaning and sanitation activities are legally required for most food operations and contribute to mitigation of pathogen transmission and recall events. Specifically, food establishments are expected to institute safe and effective cleaning and sanitation procedures to minimize the spread of pathogens detailed within sanitation standard operating procedures (SSOPs) and/or sanitation preventive controls. Regardless of the specific regulation(s) a facility/operation is subject to, keeping appropriate records of training and cleaning and sanitation practices is important for public health protection.

Although cleaning and sanitation activities are routine within the food industry, during the COVID-19 pandemic, food and agriculture industries received additional guidance from local, state, and federal officials to enhance sanitation activities (e.g., "deep clean"). Whereas, pre-COVID-19 cleaning and sanitation activities focused on reducing or eliminating foodborne pathogens of concern, which affected the end-user and not food and agriculture workers, the additional COVID-19 guidance focused on the need for disinfectants to reduce or eliminate a respiratory virus. Similar to the application of other sanitizers and disinfectants, it is important for workers to read the label and determine if these products can be applied in a food setting and if additional steps (i.e., rinse after application) are needed.

Mixed messaging causing confusion

To communicate preventive measures and enhanced cleaning and sanitation procedures to food and agriculture industry workers, as well as consumers, many organizations and institutions created multimodal (e.g., graphics, webinars, videos) and accessible (e.g., icons versus text, multilingual, nontechnical, simple language) public health communications to educate workers and consumers about the importance of personal hygiene, food safety, and cleaning and sanitation practices to bring awareness to COVID-19 preventive measures.

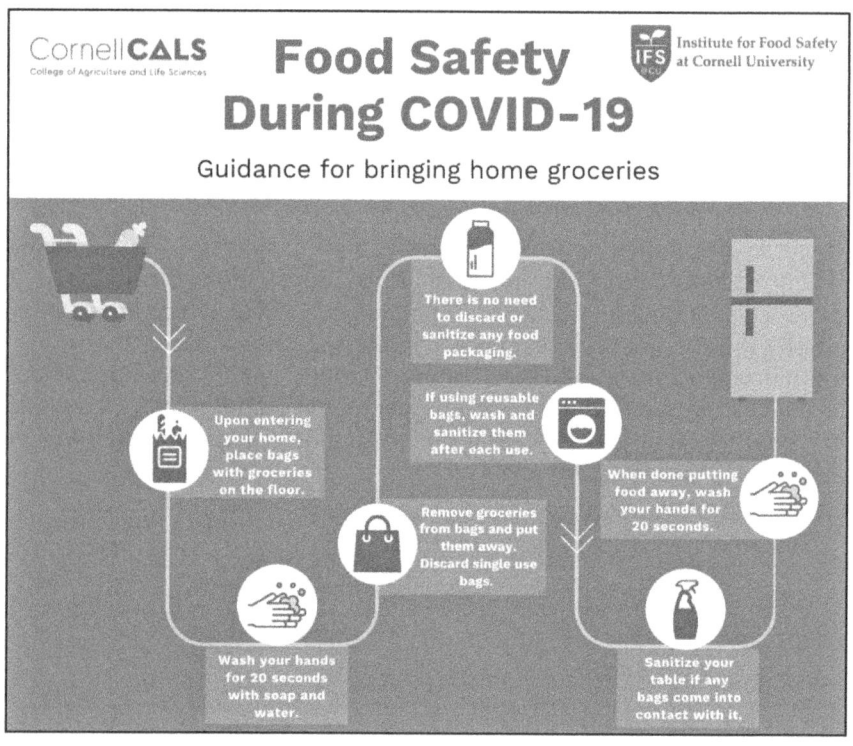

During the pandemic, U.S. public health officials also attempted to disseminate messages to communities and businesses on safe and effective disinfectant use. Despite their infographics and messaging for safer chemical application methods, the National Poison Data System (NPDS) reported substantial increases in reported exposures and poisonings from pesticides beginning in the early months of the pandemic. This could be attributed to the contradicting and false messages pertaining to cleaning products, sanitizers, and disinfectants that continued to be circulated. Examples include: (1) businesses and products falsely claiming disinfectant effectiveness against SARS-CoV-2 (U.S. Food and Drug Administration, 2020b); (2) widely circulated videos promoting the idea of applying disinfectants to food and food packaging (VanWingen, 2020); and (3) businesses encouraging their employees to walk through a mist of disinfectant to showcase purported precautionary actions to the public (Denver Broncos, 2020). These three examples, among other promotions and activities, subliminally signaled to the end-user alternative and nonapproved ways to handle disinfectants (Biswal et al., 2020).

COVID-19 AND FOOD SAFETY FAQ
IS CORONAVIRUS A FOOD SAFETY ISSUE?

CDC and USDA are not aware of any reports at this time of human illnesses that suggest COVID-19 can be transmitted by food or food packaging. However, it is always important to follow good hygiene practices (i.e., wash hands and surfaces often, separate raw meat from other foods, cook to the right temperature, and refrigerate foods promptly) when handling or preparing foods.

IS FOOD IMPORTED FROM COUNTRIES AND STATES AFFECTED BY COVID-19 AT RISK OF SPREADING COVID-19?

- Currently, there is no evidence to support transmission of COVID-19 associated with imported goods and there are no reported cases of COVID-19 in the United States associated with imported goods.

IF AN EMPLOYEE AT A FOOD ESTABLISHMENT BECAME INFECTED WITH CORONAVIRUS, WOULD THE FOOD PRODUCED AT THAT FACILITY BE SAFE TO EAT?

- Food establishment personnel who are ill with COVID-19 or any other illness should be excluded from work activities that could create unsanitary conditions (i.e. coughing or sneezing on product).
- COVID-19 is thought to spread mainly from person to person through respiratory droplets that can land in the mouths or noses of people who are nearby.

CAN I GET SICK WITH COVID-19 FROM TOUCHING FOOD, THE FOOD PACKAGING, OR FOOD CONTACT SURFACES, IF THE CORONAVIRUS WAS PRESENT ON IT?

- Currently there is no evidence of food or food packaging being associated with transmission of COVID-19.
- Coronaviruses need a living host (animal or human) to grow in and cannot grow in food.
- Like other viruses, it is possible that the virus that causes COVID-19 can survive on surfaces or objects.

HOW SHOULD FOOD BE HANDLED DURING THE COVID-19 PANDEMIC?

- As always, follow good hygiene and food safety practices when preparing food:
    - Purchase food from reputable sources
    - Cook food thoroughly and maintain safe holding temperatures
    - Use good personal hygiene
    - Clean and sanitize surfaces and equipment

NC STATE
EXTENSION

Stay informed: go.ncsu.edu/covid-19
usda.gov/coronavirus | fsai.ie/faq/coronavirus.html
Updated March 17, 2020

NC STATE
UNIVERSITY

Food and agriculture workers, government regulators, and others working in the food system have all been witness to these contradicting and false messages. With access to widespread misinformation, many food and agriculture industries have even considered implementing some non-approved methods, in an effort to keep workers safe. Another result of messaging and news from this pandemic, is that food and agriculture workers (who are also consumers) and the general consumer population have all become more aware of their personal roles/responsibilities to enact personal hygienic measures and sanitation practices in public spaces (Accenture, 2020). Consumers who had previously not been engaged in routine cleaning and sanitation activities or personal hygiene practices found themselves responsible for sanitation practices both in their homes, work (e.g., schools and offices spaces), and community spaces (e.g., churches, community gardens, gyms) (U.S. Centers for Disease Control and Prevention, 2020l).

Along with mixed messages, fraudulent products making COVID-19 claims were also adding to consumer confusion. At the time of this publication, over 100 warning letters from the FDA have been sent to companies for selling products that have not been proven effective in the prevention or treatment of COVID-19 (U.S. Food and Drug Administration, 2020b). However, significant resources (e.g., time and labor) are needed to investigate each of these potentially fraudulent products. Coupled with countless ways to purchase these products, including websites, stores, and social media, these warning letters only represent a fraction of the problem. This is comparable to how, even without a pandemic, only a certain percentage of food and agriculture facilities, depending on their size and risk level, are inspected by the FDA annually, due to lack of resources.

Consumer confusion stemming from mixed messages led to serious health implications for some individuals during the pandemic. Accentuating the negative impacts of improper usage, a May 2020 CDC survey found that there were many "gaps in knowledge about safe preparation, use, and storage of cleaners and disinfectants" (U.S. Centers for Disease Control and Prevention, 2020m). The agency claimed that roughly one-third of survey respondents, "engaged in non-recommended high-risk practices with the intent of preventing SARS-CoV-2 transmission, including using bleach on food products, applying household cleaning and disinfectant products to skin, and inhaling or ingesting cleaners and disinfectants." (U.S. Centers for Disease Control and Prevention, 2020m). The CDC also stated (U.S. Centers for Disease Control and Prevention, 2020n):

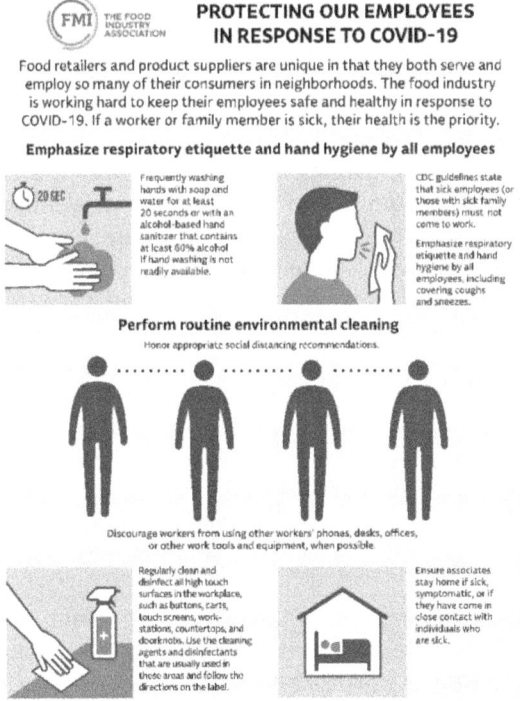

Exposures to cleaners and disinfectants reported to NPDS increased substantially in early March 2020. Associated with increased use of cleaners and disinfectants is the possibility of improper use, such as using more than directed on the label, mixing multiple chemical products together, not wearing protective gear, and applying in poorly ventilated areas. To reduce improper use and prevent unnecessary chemical exposures, users should always read and follow directions on the label, only use water at room temperature for dilution (unless stated otherwise on the label), avoid mixing chemical products, wear eye and skin protection, ensure adequate ventilation, and store chemicals out of the reach of children.

Both the FDA and the CDC websites featured first-hand accounts of pesticide misuse within the first few months of the COVID-19 response, showcasing the dangers of improper management and handling of sanitizers and disinfectants.

CDC: [The agency] reported [60%] more frequent home cleaning or disinfection compared with that in preceding months. Thirty-nine percent reported intentionally engaging in at least one high-risk practice not recommended by CDC for prevention of SARS-CoV-2 transmission (2), including application of bleach to food items (e.g., fruits and vegetables) (19%); use of household cleaning and disinfectant products on hands or skin (18%); misting the body with a cleaning or disinfectant spray (10%); inhalation of vapors from household cleaners or disinfectants (6%); and drinking or gargling diluted bleach solutions, soapy water, and other cleaning and disinfectant solutions (4% each).

(U.S. Centers for Disease Control and Prevention, 2020m)

FDA: In one recent example of consumer confusion, the FDA received a report that a consumer purchased a bottle they thought to be drinking water but was in fact hand sanitizer. The agency also received a report from a retailer about a hand sanitizer product marketed with cartoons

> for children that was in a pouch that resembles a snack. Drinking only a small amount of hand sanitizer is potentially lethal to a young child, who may be attracted by a pleasant smell or brightly colored bottle of hand sanitizer.
> *(U.S. Food and Drug Administration, 2020c)*

Shared responsibility and commitment for health in the context of the food system

Although SARS-CoV-2 does not cause foodborne illness, its impact upon the food and agriculture industry, specifically with essential workers, has highlighted the need for collaboration, outreach, and dedicated resources between all food system stakeholders to ensure the health and safety of individuals within the farm-to-fork continuum. While personal hygiene behaviors and cleaning and sanitation activities are undeniably critical for public health protection, there is a need for cultivating a culture of shared responsibility and commitment for health, worker, and food safety among government, industry, academia, and the general public to reduce and/or prevent harm to human health (UN, 2020b).

Ultimately, while COVID-19 is not a foodborne illness, it plays an obvious role in safety, as it relates to food contexts. In the context of food safety and shared responsibility, the WHO defines "shared responsibility" to include (WHO, 2020d):

> Food safety is a shared responsibility between governments, industry, producers, academia, and consumers. Everyone has a role to play. Achieving food safety is a multisectoral effort requiring expertise from a range of different disciplines – toxicology, microbiology, parasitology, nutrition, health economics, and human and veterinary medicine. Local communities, women's groups and school education also play an important role.

When expanded to the context of food and agriculture worker health and safety shared responsibilities, there is increasing need to address aspects to include:

- Ensuring essential workers have the resources to be properly trained and educated in food safety-related tasks.
- Providing paid sick leave that does not penalize the essential worker for being sick/missing work due to illness.
- Providing reliable and accessible information by ensuring and investing in health and media literacy to "meet people where they

Food safety is not just for one population, demographic, ethnicity or race. However, marginalized populations have been underserved and underrepresented by the food safety community for a variety of reasons both overt and subtle.

We do not want to sit on the sidelines waiting for others to address this issue, instead we want to be part of the change.

THE SAFE PLATES TEAM

The Safe Plates Program at NC State University is making a shared commitment to the following:

We will look at food safety through lenses of equality, equity, diversity and inclusion. We will do this to ensure that the projects we conduct, the materials we create, the teams we work with and the professionals we mentor are from and for diverse populations, backgrounds and perspectives.

We will see our work as integral to social justice and weave the needs of underserved populations into our work in research, teaching, extension and outreach.

We will provide leadership to demonstrate active participation in the dialogue and will encourage and equip our team members and colleagues to do the same.

We will communicate our commitment with all current and future partners and hold our partners to this value standard.

THE SAFE PLATES TEAM

are" so that individuals have the capacity to obtain, process, and understand basic health and media information to make appropriate decisions.

- Providing a living wage.
- Reducing food insecurity.
- Ensuring potable and safe water for all individuals.
- Increasing opportunities and access to healthcare.
- Increasing access to affordable and safe housing accommodations.
- Implementing and expanding upon practices in food safety through lenses of equality, equity, diversity, and inclusion with the goal of reaching social justice.

The COVID-19 pandemic has brought awareness to the shared responsibility in implementing personal hygiene and cleaning and sanitation practices through behavior changes in personal and workplace settings. In the food and agriculture system, respect and care for worker, consumer, and food safety has reached a crossroads where innovative and holistic approaches must be implemented to continuously improve and redefine the food system, and the concept of safety. By maintaining a shared responsibility for improving health outcomes (e.g., reducing preventable illnesses and deaths) among societies most vulnerable populations, and applying lessons learned from the pandemic, future generations can help to continue building a more sustainable, transparent and equitable food system.

References

Accenture. (2020, April 28). "COVID-19: How consumer behavior will be changed". Retrieved from https://www.accenture.com/us-en/insights/consumer-goods-services/coronavirus-consumer-behavior-research

Amnesty International. (2020, September 3). "Global: Amnesty analysis reveals over 7,000 health workers have died from COVID-19". Retrieved from https://www.amnesty.org/en/latest/news/2020/09/amnesty-analysis-7000-health-workers-have-died-from-covid19/

Arons M., Hatfield K. et al. (2020, April 24). "Presymptomatic SARS-CoV-2 infections and transmission in a skilled nursing facility". *The New England Journal of Medicine.* Retrieved from https://www.nejm.org/doi/full/10.1056/NEJMoa2008457

Baciu A., Negussie Y., Geller A., et al., (2017, January 11). "Communities in action: Pathways to health equity - the root causes of health inequity". *National Academies Press (US).* Retrieved from https://www.ncbi.nlm.nih.gov/books/NBK425845/

Biswal M., Kanaujia R., Angrup A., Ray P., Mohan Singh S. (2020, May 6). "Disinfection tunnels: potentially counterproductive in the context of a prolonged pandemic of COVID-19". *Public Health.* Retrieved from https://www.ncbi.nlm.nih.gov/pmc/articles/PMC7200329/

Braveman P. (2006, April 21). "Health disparities and health equities: Concepts and measurements". *Annual Review of Public Health.* Vol. 27:167-194. Retrieved from https://www.annualreviews.org/doi/abs/10.1146/annurev.publhealth.27.021405.102103

Center for American Progress. (2020, April 23). "Protecting farmworkers from coronavirus and securing the food supply". Retrieved from https://www.americanprogress.org/issues/economy/reports/2020/04/23/483488/protecting-farmworkers-coronavirus-securing-food-supply/

CUNY Graduate School of Public Health and Health Policy. (2020, April 23). "COVID-19 could cost the United States billions in medical expenses". *ScienceDaily.* Retrieved from https://www.sciencedaily.com/releases/2020/04/200423160512.htm

Denver Broncos. (2020, August 3). "Time for work. But first, we sanitize". *Twitter.* Retrieved from https://twitter.com/Broncos/status/1290350010547093504?ref_src=twsrc%5Etfw%7Ctwcamp%5Etweetembed%7Ctwterm%5E1290350010547093504%7Ctwgr%5Eshare_3&ref_url=https%3A%2F%2Fwww.foxnews.com%2Fsports%2Fdenver-broncos-misting-booth-coronavirus

Economic Policy Institute. (2020, May 19). "Who are essential workers? A comprehensive look at their wages, demographics, and unionization rates". Retrieved from https://www.epi.org/blog/who-are-essential-workers-a-comprehensive-look-at-their-wages-demographics-and-unionization-rates/

European Commission. (2020). "A vulnerable workforce: Migrant workers in the COVID-19 pandemic". Retrieved from https://ec.europa.eu/jrc/en/publication/vulnerable-workforce-migrant-workers-covid-19-pandemic

Food and Beverage Issue Alliance. (2020). "Important FBIA Resources". Retrieved from https://www.feedingus.org/

Goyal M., Simpson J., Boyle, M., Badolaato G., Delaney M., McCarter R., Cora-Bramble D. (2020, October 1). *Pediatrics.* Vol. 146, Issue 4. Retrieved from https://pediatrics.aappublications.org/content/146/4/e2020009951

International Commission on Microbiological Specifications for Foods. (2020, September 3). "ICMSF opinion on SARS-CoV-2 and its relationship to food safety". Retrieved from https://www.icmsf.org/wp-content/uploads/2020/09/ICMSF2020-Letterhead-COVID-19-opinion-final-03-Sept-2020.BF_.pdf

Johns Hopkins University and Medicine. (2020, September 7). "Coronavirus Resource Center". Retrieved from https://coronavirus.jhu.edu/

Khazanchi R., Evans C., Marcelin J. (2020, September 25). "Racism, Not Race, Drives Inequity across the COVID-19 Continuum". *JAMA Network Open.* Retrieved from https://jamanetwork.com/searchresults?author=Rohan+Khazanchi&q=Rohan+Khazanchi

Kim H.N., Lan K.F., Nkyekyer E., et al. (2020, September 24). "Assessment of Disparities in COVID-19 Testing and Infection across Language Groups in Seattle, Washington". *JAMA Network Open.* Retrieved from https://jamanetwork.com/journals/jamanetworkopen/fullarticle/2770951

Lancet. (2020). "The plight of essential workers during the COVID-19 pandemic. *Lancet.* Retrieved from https://www.ncbi.nlm.nih.gov/pmc/articles/PMC7241973/

Meselson M. (2020, April 15). "Droplets and Aerosols in the Transmission of SARS-CoV-2". *New England Journal of Medicine.* Retrieved from https://www.nejm.org/doi/full/10.1056/nejmc2009324

New York State. (2020, March 20). "Governor Cuomo Issues Guidance on Essential Services under the 'New York State on PAUSE' Executive Order". Retrieved from https://www.governor.ny.gov/news/governor-cuomo-issues-guidance-essential-services-under-new-york-state-pause-executive-order

North Dakota Department of H ealth. (2020). "COVID-19 glossary of terms". Retrieved from https://www.health.nd.gov/diseases-conditions/coronavirus/covid-19-glossary-terms

Price L., Nattinger A., Rivera F., et al. (2020, September 25). "Racial Disparities in Incidence and Outcomes Among Patients With COVID-19". *JAMA Network Open*. Retrieved from https://jamanetwork.com/journals/jamanetworkopen/fullarticle/2770961

Rentsch C.T., Kidwai-Khan F., Tate J.P., et al. (2020, September 22). "Patterns of COVID-19 testing and mortality by race and ethnicity among United States veterans: A nationwide cohort study". *PLoS Medicine*. Retrieved from https://journals.plos.org/plosmedicine/article?id=10.1371/journal.pmed.1003379

Stone D., Kovacevic J., Brown S.R.B. (2020, October). "Sanitizer Basics for the Food Industry. Pacific Northwest Extension Publication no. 752". Retrieved from https://catalog.extension.oregonstate.edu/pnw752.

United Nations. (2020a). "COVID-19 and Indigenous Peoples". Retrieved from https://www.un.org/development/desa/indigenouspeoples/covid-19.html

United Nations. (2020b, June 7). "World Food Safety Day: From planting to your plate, everyone has a role to play". Retrieved from https://news.un.org/en/story/2020/06/1065812#:~:text=Joining%20forces%2C%20the%20Food%20and,to%20business%20operators%20and%20consumers.

United States Department of Agriculture Economic Research Service. (2020, April 22). "Farm Labor". Retrieved from https://www.ers.usda.gov/topics/farm-economy/farm-labor/

U.S. Centers for Disease Control and Prevention. (2015, January 14). "Engineering Controls". Retrieved from https://www.cdc.gov/niosh/engcontrols/default.html#:~:text=Engineering%20controls%20protect%20workers%20by,guards%20to%20shield%20the%20worker.

U.S. Centers for Disease Control and Prevention. (2020a, May 13). "Symptoms of Coronavirus". Retrieved from https://www.cdc.gov/coronavirus/2019-ncov/symptoms-testing/symptoms.html

U.S. Centers for Disease Control and Prevention. (2020b, August 14). "People with Certain Medical Conditions". Retrieved from https://www.cdc.gov/coronavirus/2019-ncov/need-extra-precautions/people-with-medical-conditions.html

U.S. Centers for Disease Control and Prevention. (2020c, September 10). "COVID-19 Pandemic Planning Scenarios". Retrieved from https://www.cdc.gov/coronavirus/2019-ncov/hcp/planning-scenarios.html

U.S. Centers for Disease Control and Prevention. (2020d, August 21). "CDC COVID-19 Response Health Equity Strategy: Accelerating Progress Towards Reducing COVID-19 Disparities and Achieving Health Equity". Retrieved from https://www.cdc.gov/coronavirus/2019-ncov/community/health-equity/cdc-strategy.html

U.S. Centers for Disease Control and Prevention. (2020e, September 16). "Long-Term Effects of COVID-19". Retrieved from https://www.cdc.gov/coronavirus/2019-ncov/long-term-effects.html

U.S. Centers for Disease Control and Prevention. (2020f, September 25). "Update: Characteristics of Health Care Personnel with COVID-19 — United States, February 12–July 16, 2020". *MMWR*. Retrieved from https://www.cdc.gov/mmwr/volumes/69/wr/mm6938a3.htm

U.S. Centers for Disease Control and Prevention. (2020g, July 24). "Health Equity Considerations and Racial and Ethnic Minority Groups". Retrieved from https://www.cdc.gov/coronavirus/2019-ncov/community/health-equity/race-ethnicity.html

U.S. Centers for Disease Control and Prevention. (2020h, August 18). "COVID-19 Hospitalization and Death by Race/Ethnicity". Retrieved from https://www.cdc.gov/coronavirus/2019-ncov/covid-data/investigations-discovery/hospitalization-death-by-race-ethnicity.html

U.S. Centers for Disease Control and Prevention. (2020i, August 10). "Criteria for Return to Work for Healthcare Personnel with SARS-CoV-2 Infection (Interim Guidance)". Retrieved from https://www.cdc.gov/coronavirus/2019-ncov/hcp/return-to-work.html

U.S. Centers for Disease Control and Prevention. (2020j, August 22). "Food and Coronavirus Disease 2019 (COVID-19)". Retrieved from https://www.cdc.gov/coronavirus/2019-ncov/daily-life-coping/food-and-COVID-19.html#:~:text=Currently%2C%20there%20is%20no%20evidence,coughs%2C%20sneezes%2C%20or%20talks

U.S. Centers for Disease Control and Prevention. (2020k, April 13). "Key Definitions & Abbreviations". Retrieved from https://www.cdc.gov/hai/prevent/resource-limited/definitions.html#anchor_1585234497534

U.S. Centers for Disease Control and Prevention. (2020l, September 10). "Cleaning and Disinfection for Community Facilities". Retrieved from https://www.cdc.gov/coronavirus/2019-ncov/community/organizations/cleaning-disinfection.html

U.S. Centers for Disease Control and Prevention. (2020m, June 11). "Knowledge and Practices Regarding Safe Household Cleaning and Disinfection for COVID-19 Prevention — United States, May 2020". Retrieved from https://www.cdc.gov/mmwr/volumes/69/wr/mm6923e2.htm

U.S. Centers for Disease Control and Prevention. (2020n, April 24). "Cleaning and Disinfectant Chemical Exposures and Temporal Associations with COVID-19 — National Poison Data System, United States, January 1, 2020–March 31, 2020". Retrieved from https://www.cdc.gov/mmwr/volumes/69/wr/mm6916e1.htm

U.S. Cybersecurity and Infrastructure Security Agency. (2020, August 18). "Identifying Critical Infrastructure During COVID-19". Retrieved from https://www.cisa.gov/identifying-critical-infrastructure-during-covid-19

U.S. Environmental Protection Agency. (2020a, August 11). "What's the difference between products that disinfect, sanitize, and clean surfaces?" Retrieved from https://www.epa.gov/coronavirus/whats-difference-between-products-disinfect-sanitize-and-clean-surfaces

U.S. Environmental Protection Agency. (2020b, August 12). "Determining If a Cleaning Product Is a Pesticide under FIFRA". Retrieved from https://www.epa.gov/pesticide-registration/determining-if-cleaning-product-pesticide-under-fifra

U.S. Environmental Protection Agency. (2020c, July 7). "Disinfectant Use and Coronavirus (COVID-19)". Retrieved from https://www.epa.gov/coronavirus/disinfectant-use-and-coronavirus-covid-19

U.S. Environmental Protection Agency. (2020d, October 5). "List N: Disinfectants for Coronavirus (COVID-19)". Retrieved from https://www.epa.gov/pesticide-registration/list-n-disinfectants-coronavirus-covid-19

U.S. Environmental Protection Agency. (2020e, August 11). "How does EPA know that the products on List N work on SARS-CoV-2?" Retrieved from https://www.epa.gov/coronavirus/how-does-epa-know-products-list-n-work-sars-cov-2

U.S. Environmental Protection Agency. (2020f, August 11). "Will EPA take enforcement action against companies making false claims that their disinfectants work against SARS-CoV-2 (COVID-19)?" Retrieved from https://www.epa.gov/coronavirus/will-epa-take-enforcement-action-against-companies-making-false-claims-their

U.S. Environmental Protection Agency. (2020g, August 11). "I can't tell if the product I'm interested in is on the list or not. Can you help me?" Retrieved from https://www.epa.gov/coronavirus/i-cant-tell-if-product-im-interested-list-or-not-can-you-help-me

U.S. Environmental Protection Agency. (2020h, August 11). "I have a question about a word or phrase on the List N website. I'm not sure how something on List N helps me fight COVID-19." Retrieved from https://www.epa.gov/coronavirus/i-have-question-about-word-or-phrase-list-n-website-im-not-sure-how-something-list-n

U.S. Environmental Protection Agency. (2020i, August 11). "I want to use a product to kill SARS-CoV-2 (COVID-19) but it isn't on List N. Is it effective against SARS-CoV-2 (COVID-19)?" Retrieved from https://www.epa.gov/coronavirus/i-want-use-product-kill-sars-cov-2-covid-19-it-isnt-list-n-it-effective-against-sars-cov

U.S. Food and Drug Administration. (2020a, May 22). "Food and Agriculture: Considerations for Prioritization of PPE, Cloth Face Coverings, Disinfectants, and Sanitation Supplies During the COVID-19 Pandemic". Retrieved from https://www.fda.gov/food/food-safety-during-emergencies/food-and-agriculture-considerations-prioritization-ppe-cloth-face-coverings-disin fectants-and

U.S. Food and Drug Administration. (2020b, October 8). "Fraudulent Coronavirus Disease 2019 (COVID-19) Products". Retrieved from https://www.fda.gov/consumers/health-fraud-scams/fraudulent-coronavirus-disease-2019-covid-19-products

U.S. Food and Drug Administration. (2020c, August 27). "COVID-19 Update: FDA Warns Consumers about Hand Sanitizer Packaged in Food and Drink Containers". Retrieved from https://www.fda.gov/news-events/press-announcements/covid-19-update-fda-warns-consumers-about-hand-sanitizer-packaged-food-and-drink-containers

World Health Organization. (2020a). "Naming the coronavirus disease (COVID-19) and the virus that causes it". Retrieved from https://www.who.int/emergencies/diseases/novel-coronavirus-2019/technical-guidance/naming-the-coronavirus-disease-(covid-2019)-and-the-virus-that-causes-it

World Health Organization. (2020b, September 7). "Coronavirus disease (COVID-19) pandemic". Retrieved from https://www.who.int/emergencies/diseases/novel-coronavirus-2019

World Health Organization. (2020c, July). "Water, sanitation, hygiene, and waste management for SARS-CoV-2, the virus that causes COVID-19". Retrieved from https://apps.who.int/iris/bitstream/handle/10665/333560/WHO-2019-nCoV-IPC_WASH-2020.4-eng.pdf?ua=1

World Health Organization. (2020d, June 7). "World Food Safety Day 2020". Retrieved from https://www.who.int/news-room/events/detail/2020/06/07/default-calendar/world-food-safety-day-2020#:~:text=Food%20safety%20is%20a%20shared,cause%20damages%20to%20our%20health

VanWingen J. (2020, March 24). "PSA Grocery Shopping Tips in COVID-19 (See Important Notes Below) www.DrJeffVW.com". *YouTube*. Retrieved from https://www.youtube.com/watch?v=sjDuwc9KBps

chapter seven

Fruits from agroforestry as sustainable meat and vegetable replacements: The case of the Artocarpus

Paul De Filippi

Contents

EDITORS' NOTE: EMERGING PLANT-BASED NUTRIENTS AND STAPLE CROP RESILIENCE

To promote intersections between climate change policy, environmental resiliency, supply chain efficiency, and nutrition, it is important to consume a greater variety of plant-based foods, especially from local sources. Plant-based foods are thought to be more sustainable and energy-efficient than industrial-scale animal-based farming practices. The Good Food Institute, a nonprofit organization that promotes plant-based foods, states that a massive consumer shift in plant-based commodities can help reduce (1) land use; (2) greenhouse gases; (3) antibiotic use on livestock; and (4) water use. Recently, a group of scholars published a landmark work on succulent plant-based diets entitled *Plant-Based Diets for Succulence and Sustainability*.[1] Here, the editor, Kathleen Kevany, "takes an interdisciplinary look at how the transformation towards plant-based diets is becoming more culturally acceptable, economically accessible, technically available and politically viable."[2]

Although from a food safety perspective, produce farms in the United States that are in close proximity to small and large concentrated animal feeding operations (CAFOs) have also been implicated in major food safety outbreaks. Local land-use and environmental laws prevent certain foodborne outbreak investigators from identifying and characterizing common pathogens on CAFOs that may contaminate and implicate nearby produce farms with genetically similar bacterial pathogens derived from animal fecal sources. Birds, pests, and workers also have the potential to contaminate plant-based foods through the introduction of common food safety hazards also found in fecal material. As genetic subtyping tools, like whole genome sequencing, continue to be implemented by regulators at local levels and uploaded into collaborative international databases, opportunities exist to improve food safety practices on plant-based farms throughout the world. Therefore, stakeholders that grow, produce, and sell plant-based food products must account for food safety and ensure that their products are not adulterated or misbranded along with every other agricultural commodity.

Magnifying the potential of plant-based diets and the advantages to a more controlled food system through improved transparency are at the heart of this book. Plant-based diets may help play a role, as the following chapter illustrates, in increasing accountability and reducing the haziness of the modern food system. In the subsequent chapter, Paul De Filippi, who grows jackfruit, breadfruit, figs, and other plant-based crops in Hawaii, has a deep understanding of the regulatory aspects of the food systems and shares benefits of plant-based foods to help improve environmental and social well-being at local and international levels.

[1] https://www.taylorfrancis.com/books/e/9780429427138
[2] https://www.taylorfrancis.com/books/e/9780429427138

Making plant-based food mainstream

Tropical fruit tree crops have the potential to expand into mainstream food production as both vegetable and meat replacements. Common examples of replacement products often include industrially farmed soy-based proteins, such as tofu, or wheat-based seitan. Replacement products derived from fruit crops are exemplified by the genus *Artocarpus* under the family *Moraceae* (Paull & Duarte, 2012). The genus is best known for the jackfruit and breadfruit species. Both jackfruit and breadfruit have fostered emerging international markets for numerous fresh and processed food products (Thillakawardane, 2009). Moreover, both of these fruits are touted as potential means to address world hunger through their current underutilized abundant productions, lack of pests, low inputs, speed of production, long crop-life, ability to withstand climate change, minimal required processing steps, and potentially sustainable productions (APAARI, 2012). However, these fruits and their niche products have a limited ability to penetrate the mainstream food market and compete with more well-known, cheaper, staple foods. "Multistakeholder initiatives" are required to raise standards of food production, consumer expectations, awareness, and allow these sustainable products to become more mainstream (Smith, 2008, p. 859). The advantages of improving the sustainability of the food supply chain are countless, but this chapter summarizes just a few from the perspectives of production and marketing, geographic distribution, environmental, and socioeconomic impacts.

Nutrient Profile (per 100 g)*						
Nutrient	Jackfruit	Jackfruit Seeds	Breadfruit	Rice	Potato	Beef
Calcium (mg)	22-37	0.05-0.55	16.8	3	9	4.6-30.9
Carbohydrate (g)	15.1-25.4	38.4	31.9	28.6	15.7	0
Fiber (g)	1.0-5.0	1.5	5.4	0.3	2.4	0
Iron (mg)	0.5-1.7	0.002-1.2	0.5	0.2	0.5	1.4-3.2
Lutein (µg)	-	-	96.3	0	0	-
Magnesium (mg)	-	-	34.3	13	21	18-28
Niacin (mg)	0.4-4.0	-	0.9	0.4	1.06	4.9-8.6
Phosphorus (mg)	38	0.13-1.23	43.1	37	62	164-237
Potassium (mg)	292-407	-	376.7	29	407	321-365
Protein (g)	1.3-1.9	6.6	4	2.4	1.7	20-30
Riboflavin (mg)	0.06	-	0	0.016	0.03	0.12-0.15
Sodium (mg)	17564	-	19.4	0	16	50-62
ß-Carotene (µg)	-	-	15.1	0	0	-
Thiamine (mg)	0.03	-	0.1	0.02	0.07	0.06-0.09
Vitamin A (IU)	66-540	-	41	0	0	0
Vitamin C (mg)	44053	-	2.4	0	9.1	0
Zinc (mg)	-	-	0.1	0.42	0.29	3.3-5.6

* (Breadfruit Institute, 2014), (Paull & Duarte, 2012), (Love & Paull, 2011), (USDA, 2011)

Compared to more well-known foods, jackfruit and breadfruit are underutilized fruits, despite being well suited to serve as nutrient-rich ingredients and replacements in a diverse range of food products (Swami, Thakor, Haldankar & Kalse, 2012). Health benefits of the fruits include, but are not limited to their low-caloric profiles, high fiber contents, potassium, and protein. The following table illustrates the comparative nutrient profiles from common sources to jackfruit and breadfruit.

The attributes of jackfruit and breadfruit lend themselves to sustainable commercial viability, including supporting and feeding the rural, agricultural, low-income communities in productive tropical regions (Liu, Ragone & Murch, 2015). Improving aspects of production, processing, and transportation have the potential to further increase the sustainability of these food products and their supply chains.

Strong sustainable consumption approaches

Globally traded, commercially farmed foods and fruit-based products share many unsustainable impact categories with traditional meats and vegetables including carbon emissions from production and distribution, loss of biodiversity from agriculture, as well as chemical use (Röös, Ekelund & Tjärnemo, 2014). However, the unsustainable aspects of the fruit products are minor in comparison to the associated costs of the consumption of livestock products. These aspects can also be mitigated through planned sustainable growth and supply chain improvements of the immature market (APAARI, 2012). Each stage of production, processing, and distribution can be improved to increase efficiency and viability through education, promoting strategies such as intercropping, farmer cooperatives, novel processing, and packaging equipment. By comparison, commercial livestock productions do not share the same baseline sustainability nor potential improvements and straightforward fixes, such as shortened supply chains and sustainable land-use management at scale (FAO, n.d.).

The sustainability of a food product and its associated supply chain can be assessed through its effect on the health and welfare of each stakeholder group, including the public, consumers, producers, employees, and the environment (Smith, 2008). The ability of fruit products to be produced sustainably and to replace unsustainable products, such as meats, makes them suitable "strong sustainable consumption approaches," with a capacity to change and reduce consumptive patterns (Fuchs & Lorek, 2005 p. 262).

Production and marketing: Intensive farming versus agroforestry

Production and marketing of jackfruit and breadfruit are currently limited by a lack of awareness of the crops and their potential products, as well as climactic production requirements. The size variations of the fruits alone make them an intimidating venture for any consumer. Furthermore, the lack of specialized equipment suited for production and processing, makes them difficult crops to manage at scale (APAARI, 2012). As relatively immature markets, production, processing, and distribution have not been fully integrated to achieve its profit potential. Therefore, it is important to educate and invest in emerging markets in a commercially sustainable manner in order to avoid future costs for stakeholders and the environment (Van Den Berg & Jiggins, 2007).

Breadfruit and jackfruit have traditionally been grown in their indigenous regions as individual trees in diverse agroforestry systems on small farms and in homesteads, being used for subsistence and local commerce (APAARI, 2012). Depending on the production region, trees can thrive and produce fruit without irrigation, only requiring supplemental irrigation during long periods of drought and in arid climates (Love & Paull, 2011). Due to their abundance of production, limited processing, and distribution chains, there have historically been periodic surpluses with large quantities of waste and profit-loss (Sundarraj & Ranganathan, 2018). Recently, these crops have expanded their agricultural footprints and have become valuable commodities leading to more intensive production strategies and monocrop orchards (APAARI, 2012). One the one hand, this intensive farming is more specialized and can have higher output with reduced labor, which improves the producers' competitiveness. On the other hand, intensification may represent a shift away from traditional agroecology and agroforestry, creating a reliance on costly inputs and a single crop with no diversification.

According to this author's experience in cultivating jackfruit in Hawaii, adopting traditional agroforestry systems has the potential to increase the overall productivity, diversity, profitability, and sustainability of the agricultural operations (Miah et al., 2017). Utilizing a system of overstory, midstory, and understory crops can reduce fertilizer inputs and risk of crop failure, while increasing the productivity of the operation (Hegde, 2009). As an example, jackfruit can be intercropped with high-value tree and vining crops such as mango, lychee, longan, vanilla, and peppercorn. These fruits can be grown above understory crops such as ginger, turmeric, cardamom, peanuts, and sweet potato. Thus, providing growers with opportunities for collaborative agricultural education along with access to improved cultivars and local technologies has the potential to increase overall yields and extend the lifespan of their crops (Daniel et al., 2014). This increased value associated with improved cultivars and practices can help avoid trees being cut down for timber. Processing technologies that will ease the skinning and cutting of the large fruits while minimizing waste are necessary to increase efficiency and food production.

The input and land requirements of a diversified agroforestry system are in sharp contrast to those of intensive crop productions such as rice, which requires flood irrigation and results in substantial greenhouse gas emissions (Neue, 1993). Unlike traditional field crops, these fruits can be grown on otherwise marginable land including rocky soils and hillsides, as long as adequate soil moisture can be maintained. Nitrogen fixing and otherwise productive crops can be used as ground cover and intercrops, further reducing the need for chemical inputs and irrigation.

The crops can be grown domestically in the United States in Hawaii and Florida, however, factors such as limited investment, limited labor, lack of mechanization, and high cost of production prevent them from extensive commercialization. Productions in these areas must rely on high priced, niche, local markets with added value through shortened supply chains, novel products, removing intermediaries, branding, corporate social responsibility, and transparency (Smith, 2008). Regions with higher labor or land costs can be outcompeted by exporting countries with lower overall costs of production, longer supply chains, and potentially less transparency.

The market has created the need for intermediary traders and wholesalers to collect and distribute fruit from these systems. While the majority of trade networks are still local, they have opened new opportunities for food processors to create and distribute products to broader domestic and international markets (APAARI, 2012). As these systems grow, efforts must be taken on the part of the producers and the consumers to ensure that the supply chain remains transparent and free from adulteration (Smith, 2008). Each stakeholder in the supply chain has the responsibility to ensure the quality and integrity of their procurements. Further market expansion can be achieved sustainably through increased education and awareness of the fruits, their productions, products, and by-products (Bapat, Jagtap, Ghag & Ganapathi, 2019). Full utilization of the fruit crops and their expansive value-added potential can further increase their viability and future innovations. Communication and collaboration between growers both locally and internationally can help this process by opening new markets, business opportunities, and increasing global supply and demand.

Breadfruit

The breadfruit tree, known in Hawaii as ulu, can produce up to 400 kilograms of fruit per year at maturity, with individual weights varying by cultivar. In contrast with other starches such as rice and potato, the fruit contains all of the essential amino acids and boasts a protein content up to 7.6 percent by dry weight, approximately 4 grams per serving depending on the cultivar (Liu, Ragone & Murch, 2015). The breadfruit has embodied its namesake through milling the fruit into a flour product high in starch and fiber (Clark & Aramouni, 2018). The flour is currently used in niche markets for gluten-free baking and an alternative to traditional flour. The fruit is also used in its unripe stage as a general starch and can be boiled, dried, steamed, fried, or pickled (Liu, Ragone & Murch, 2015).

Breadfruit is traditionally baked, unripe in a fire, and consumed as a starch. In modern food production, it is commonly used as a substitute for potatoes in products such as potato chips or wedges, mashed potatoes, salads, curries, and doughs. The ripe fruit is not utilized to the same extent due to its limited shelf life, inability to transport, and lack of commercialization. It can be described as having a sweet tasting, pancake batter consistency ideal for baked goods.

Jackfruit

The jackfruit tree can bear up to 500 fruit per year at full maturity, with individual weights averaging approximately 16 kilograms (Love & Paull, 2011). Fruits can weigh up to 50 kilograms each, making jackfruit the largest tree fruit in the world (Bapat, Jagtap, Ghag & Ganapathi, 2019). While the jackfruit pulp contains a modest amount of protein, the total protein content per serving, including its numerous seeds, can be over 13.5 grams (Jurez-Barrientos et al., 2017). The jackfruit is similar to breadfruit in its broad range of applications. It is described as having five major uses in its various stages. The tender undersized fruit, the larger immature fruit, the mature unripe fruit, the mature ripe fruit, and finally the seeds of the fruit (Swami, Thakor, Haldankar & Kalse, 2012). The immature forms of the fruit, traditionally used as a vegetable, have been popularized as a meat substitute in Western markets. The products are commonly marketed in North America and Australia as vegan pork or chicken due to its consistency and ability to absorb fats and flavors (APAARI, 2012).

Jackfruit is traditionally eaten ripe as a sweet fruit, or cooked unripe as a vegetable with the seeds roasted and mashed into a paste depending on the culture of the region. In modern food production, jackfruit has created a niche market for itself as a meat replacement or a substantive addition to many dishes. Depending on the desired flavor profile, the texture can lend itself to a number of meat products such as pork, chicken, or fish.

Geographic distribution and agricultural benefits

The geographic distribution of jackfruit and breadfruit warrants a discussion of the agroecological benefits, such as pest-resistance and indigenous knowledge preservation. Breadfruit is native to New Guinea and the surrounding Pacific Islands. While jackfruit originated in the neighboring Indo-Malay region, both are grown and distributed throughout the surrounding areas (Paull & Duarte, 2012). Over centuries, the fruits, and their propagative plant material, have travelled with human migrations throughout Asia and the tropical Pacific, becoming an integral part of many cultures practicing subsistence farming (Liu, Ragone & Murch, 2015). Production has now spread to tropical and subtropical regions worldwide including Hawaii, Florida, the Caribbean, Mexico, South America, and parts of Africa.

Pest resistance

Pests are a serious economic issue for all agricultural crops and food productions. They can include fungus, bacteria, insects, rodents, and birds. These pests result in crop and profit losses every year (Oerke, 2005). In Hawaii and many other tropical areas, integrated pest management systems are needed in order for agricultural operations to remain sustainable, minimizing inputs, and maximizing profits (Van Den Berg & Jiggins, 2007). For example, without netting, deterrents, and proper pest management, birds can completely wipe out certain crops. Like insect and fungal pests, birds can adapt and become resistant to poorly managed treatments including physical, visual, audible, and chemical deterrents (Tabashnik et al., 1997). Insect pest pressure can be more severe in tropical areas; without cold winter temperatures to control insect populations, problematic invasive species have the potential to establish and thrive. These climates can provide ideal environments for species such as fruit flies and various moths to proliferate and inflict severe economic damage on a multitude of crops. Pesticides have long been the favored method of control by the agricultural industry because of their effectivity and efficiency, often decreasing the number of hours and employees needed to maintain a crop in the short-term (Popp, Pető & Nagy, 2012).

Pesticide-resistance presents a major challenge for global agriculture, increasing the quantities of pesticides being applied and precipitating an ongoing need for new chemical formulations (Tabashnik et al., 1997). Even though the U.S. Environmental Protection Agency (EPA) approves hundreds of pesticides, the use of these chemicals, many of which are synthetic, can endanger sustainable agriculture. The diamondback moth, which is a pest of brassica crops, is a prime example of how costly an invasive species can be, and how pest-resistance can exacerbate pest problems and limit treatment options. The moth has become resistant to several common insecticides which has led to the widespread adoption of a more involved and costly resistance management systems. These systems utilize increased labor, rotations of pesticides and crops, and the introductions of various hopeful biological controls such as bacteria and parasites (Tabashnik et al., 1997).

Within the international plant trade industry, there are always concerns with the introduction of plant pest species. However, *Artocarpus's* slow growth nature and limited ability to self-propagate allow the production of this species to be easily controlled and monitored (APAARI, 2012). These species boast dense canopies which restrict light and reduce the need for weeding or herbicide applications (Love & Paull, 2011). Additionally, the thick exterior of the fruit and methods of processing limit the potential negative effects of many insect pests. While some insect, bacterial, and fungal damage can occur on fruits, the significant issues arise from fungal contamination of the roots which can significantly affect the lifespan of the tree (Love & Paull, 2011). This can be avoided through good agricultural practices to monitor and limit exposure to contaminated materials and avoid planting in areas without sufficient drainage. As a result, the reduced need for chemical inputs make jackfruit and breadfruit excellent agricultural choices when paired with sustainable, integrated pest management and agroforestry systems (Oerke, 2005).

Indigenous knowledge preservation

The specific climactic requirements of the *Artocarpus* genus, paired with cultural knowledge, present a unique opportunity for indigenous populations in producing areas to capitalize on these versatile crops (APAARI, 2012). Educating producers to maintain and expand traditional agricultural techniques while investing in local innovations can help to preserve the cultural aspects of the products while decreasing the reliance on chemical inputs (Van Den Berg & Jiggins, 2007).

The productive areas of India have organized and created distribution systems capable of supplying jackfruit to less productive states where jackfruit is not commonly grown or used in the same way. Similarly, growers in Hawaii have begun to form cooperatives for processing to take advantage of the benefits of scale, while also promoting the cultural importance and heritage of the Polynesian fruit that has been cultivated in the region for thousands of years (Hawai'i 'Ulu Cooperative, 2020). The diversity of growers and cultures worldwide allows for specific and unique applications and replacement potential for these fruit products. This has also allowed motivated stakeholders to expand or in some cases create agricultural markets for the fruits and their products. Countries such as Australia, China, Columbia, Indonesia, Kenya, Malaysia, Mexico, the Philippines, and Sri Lanka, along with their local industries, have created manufacturing and export markets to service international demand (APAARI, 2012). By example, the Philippines and India have created local and national initiatives, including subsidies, to increase production of jackfruit and reduce waste (Philippine Information Agency, 2019). These types of government programs, like the transportation subsidy in India, benefit the entire supply chain and allow for overall market growth (Madhvi, 2020).

The new production centers have the potential to reach local, domestic, and international consumers, decreasing the need for overseas transport, limiting the cost of shipping and growing the global market. However, new production centers also have the potential to compete with indigenous growing regions, reducing their profits and sustainability (Keen, 2015). Overall, island communities and other significant importers have the opportunity to use these crops to become more productive, self-sustainable, and decrease their reliance on foreign products (Liu, Ragone & Murch, 2015).

Environmental impacts

Utilizing sustainable agricultural practices such as agroforestry to produce replacement crops provides a multifaceted approach to improving the environmental impact of food production. By its very nature, the agroforestry model helps mitigate common environmental pitfalls that conventional agricultural models are exposed to, including pests, chemical inputs, deforestation, and unsustainable land-use management (Smith, 2008).

The production of fruit and subsequent fruit products has the potential to produce significant quantities of waste in the form of biomass (Li, Fan, Wu, Jiang & Shi, 2019). The breadfruit and the jackfruit each contain approximately 70 and 28 percent edible portions, respectively, making jackfruit the more substantial contributor to potential environmental waste, depending on cultural practice (Paull & Duarte, 2012). The waste biomass can be repurposed for additional uses including human food, animal feed, fuels, medicines, and other industrial products (APAARI, 2012). Industrial applications of products such as animal feed and fruit-derived pectin can increase the overall sustainability of any subsequent food products (Li, Fan, Wu, Jiang & Shi, 2019). By implementing agroforestry practices and recycling biological wastes, farmers can decrease reliance on chemical fertilizers and pesticides, while also mitigating environmental losses from deforestation and habitat destruction (Oliveira, Neves, Raboy & Dietz, 2010).

The majority of the demand for these fruit products is currently within the producing areas, primarily due to consumer awareness and access. However, demand from the temperate regions is growing, requiring expanded global supply chains and packaging to get products to consumers. These environmental impacts may be mitigated in the future through the utilization of fruit by-products such as biofuel (Soetardji,

Widjaja, Djojorahardjo, Soetaredjo & Ismadji, 2014). Further research is required to investigate the food safety and durability of biodegradable film packaging made from fruit starches, which would further mitigate environmental impacts (Bonomo et al., 2017).

Socioeconomic impacts

Production of these Artocarpus fruits has the potential to benefit individuals across the socioeconomic spectrum in both commercial and cultural applications (Hegde, 2009). The cultivation and uses of these fruits are not common knowledge; this presents an opportunity for localized and specialized production. Each production area can incorporate these fruits into their own culturally specific products, creating an in-demand product for domestic consumption and a novel or niche product for international markets (APAARI, 2012). The ability for production to sustain local populations in addition to generating income is particularly important in the tropics, where malnutrition and specifically protein deficiency can be prevalent. These productions have the potential to avoid illnesses and deaths of these at-risk populations (Liu, Ragone & Murch, 2015).

In the case of quinoa, another novel source of protein, worldwide popularization and intensification of production led to increases in the quality of life for populations in the producing regions of Bolivia and Peru, generating more income to invest in their local communities (Rojas-Ruiz, 2012). However, increased foreign demand has made quinoa, a South American staple, unaffordable to local consumers. Additionally, increased reliance on a single crop, utilizing cheap, potentially unsustainable methods to rapidly increase production has left these indigenous production areas at risk of being undercut, with no alternative incomes, by new production regions (Keen, 2015). The emerging production centers of Canada, Australia, and China have the potential to collapse the indigenous markets through improved technologies, efficiencies, and shortened supply-chains. This market sensitivity can be mitigated through sustainable agricultural practices focused on crop and product diversity, and maintaining a stable local market, and profitability.

In order to increase production, efficiency, utilization, and profitability, while also decreasing waste and maintaining socioeconomic sustainability, further research and market education are required (Bapat, Jagtap, Ghag & Ganapathi, 2019). Proper utilization of associated waste products can increase the profitability of the fruit and minimize disposal costs (Sundarraj & Ranganathan, 2018). As regions transition from small producers to commercial scale, wholesalers and intermediaries still reap the majority of the benefits due to the immaturity of the market and disorganization of farmer groups (APAARI, 2012). It is important for producing regions to invest in both production and processing industries in order to capture the full economic value in the supply chain and ensure fair distribution of benefits among local stakeholders.

Conclusion

Broad commercialization of the jackfruit and breadfruit markets has the potential to be conducted in a sustainable manner that benefits both global stakeholders and the environment as a viable, nutritional meat substitute. The sustainability of the products is based on standards for "quality, safety, and environmental performance," and their potential to create value for the producer, consumer, and the entire supply chain (Smith 2008, p. 852). Stakeholders and policy makers must steer industry to avoid unsustainable pitfalls that other global markets have fallen into. Supply chains must be managed to avoid disenfranchising local cultures, producers, processors, and retailers. Strategies should enable them to sustain their operations and provide for their local regions. Each stage of production and consumption should be shaped by policy and industry to ensure that it develops in a sustainable fashion that takes advantage of shortened supply chains, indigenous methodologies, novel technologies, and strategies for continuous improvement (Fuchs & Lorek, 2005).

References

APAARI. (2012). *Jackfruit Improvement in the Asia-Pacific Region – A Status Report* (p. 182, Rep.). Bangkok, Thailand: Asia-Pacific Association of Agricultural Research Institutions.

Bapat, V. A., Jagtap, U. B., Ghag, S. B., & Ganapathi, T. R. (2019). Molecular Approaches for the Improvement of Under-Researched Tropical Fruit Trees: Jackfruit, Guava, and Custard Apple. *International Journal of Fruit Science, 20*(3), 233–281. doi:10.1080/15538362.2019.1621236

Bonomo, R. C., Santos, T. A., Santos, L. S., Fontan, R. D., Rodrigues, L. B., Pires, A. C., … Bonomo, P. (2017). Effect of the Incorporation of Lysozyme on the Properties of Jackfruit Starch Films. *Journal of Polymers and the Environment, 26*(2), 508–517. doi:10.1007/s10924-016-0902-4

Breadfruit Institute. (2014). Breadfruit Marketing Materials. Retrieved September 08, 2020, from https://hdoa.hawaii.gov/add/main/ulumaterials/

Clark, E. A., & Aramouni, F. M. (2018). Evaluation of Quality Parameters in Gluten-Free Bread Formulated with Breadfruit (*Artocarpus altilis*) Flour. *Journal of Food Quality, 2018*, 1–12. doi:10.1155/2018/1063502

Daniel, R., Borines, L. M., Soguilon, C., Montiel, C., Palermo, V. G., Guadalquiver, G. A., … Guest, D. (2014). Development of Disease Management Recommendations for the Durian and Jackfruit Industries in the Philippines using Farmer Participatory Research. *Food Security, 6*(3), 411–422. doi:10.1007/s12571-014-0352-6

FAO. (n.d.). Cattle Ranching and Deforestation. Retrieved September 12, 2020, from http://www.fao.org/3/a-a0262e.pdf

Fuchs, D. A., & Lorek, S. (2005). Sustainable Consumption Governance: A History of Promises and Failures. *Journal of Consumer Policy, 28*(3), 261–288. doi:10.1007/s10603-005-8490-z

Hawai'i 'Ulu Cooperative. (2020). Hawai'i 'Ulu Cooperative. Retrieved September 08, 2020, from https://eatbreadfruit.com/

Hegde, N. (2009). Promotion of Underutilized Crops for Income Generation and Environmental Sustainability. *Acta Horticulturae,* (806), 563–570. doi:10.17660/actahortic.2009.806.70

Jurez-Barrientos, J., Hernndezsantos, B., Hermanlara, E., Martnezsnchez, C., Torrucouco, J., Ramrezrivera, E., … Rodrguezmiranda, J. (2017). Effects of Boiling on the Functional, Thermal and Compositional Properties of the Mexican Jackfruit (*Artocarpus heterophyllus*) Seed. *Emirates Journal of Food and Agriculture, 29*(1), 1. doi:10.9755/ejfa.2016-08-1048

Keen, T. (2015). *The Battle for Quinoa: The Bittersweet Reality of Globalization.* Washington: The Council on Hemispheric Affairs. Retrieved from ProQuest One Academic; Social Science Premium Collection Retrieved from https://link.ezproxy.neu.edu/login?url=https://www-proquest-com.ezproxy.neu.edu/docview/1667731506?accountid=12826

Li, W., Fan, Z., Wu, Y., Jiang, Z., & Shi, R. (2019). Eco-friendly Extraction and Physicochemical Properties of Pectin from Jackfruit Peel Waste with Subcritical Water. *Journal of the Science of Food and Agriculture, 99*(12), 5283–5292. doi:10.1002/jsfa.9729

Liu, Y., Ragone, D., & Murch, S. J. (2015). Breadfruit (*Artocarpus altilis*): A Source of High-quality Protein for Food Security and Novel Food Products. *Amino Acids, 47*(4), 847–856. doi:10.1007/s00726-015-1914-4

Love, K., & Paull, R. E. (2011). Jackfruit. *Fruits and Nuts,* f_n-19. Retrieved August 15, 2020, from https://scholarspace.manoa.hawaii.edu/bitstream/10125/33296/F_N-19.pdf

Madhvi, S. (2020). Govt to Give 50% Subsidy for Fruit, Vegetable Transport to Help Farmers Cut Post-harvest Loss, Avoid Distress Sale. Retrieved September 8, 2020, from https://economictimes.indiatimes.com/news/economy/agriculture/govt-to-give-50-subsidy-for-fruit-vegetable-transport-to-help-farmers-cut-post-harvest-loss-avoid-distress-sale/articleshow/76339679.cms?from=mdr

Miah, M. G., Islam, M. M., Rahman, M. A., Ahamed, T., Islam, M. R., & Jose, S. (2017). Transformation of Jackfruit (*Artocarpus heterophyllus* Lam.) Orchard into Multistory Agroforestry Increases System Productivity. *Agroforestry Systems, 92*(6), 1687–1697. doi:10.1007/s10457-017-0118-1

Neue, H. (1993). Methane Emission from Rice Fields. *BioScience, 43*(7), 466–474. doi:10.2307/1311906

Oerke, E. (2005). Crop Losses to Pests. *Journal of Agricultural Science, 144*(1), 31–43. doi:10.1017/s0021859605005708

Oliveira, L. C., Neves, L. G., Raboy, B. E., & Dietz, J. M. (2010). Abundance of Jackfruit (*Artocarpus heterophyllus*) Affects Group Characteristics and Use of Space by Golden-Headed Lion Tamarins (*Leontopithecus chrysomelas*) in Cabruca Agroforest. *Environmental Management, 48*(2), 248–262. doi:10.1007/s00267-010-9582-3

Paull, R. E., & Duarte, O. (2012). Breadfruit, Jackfruit, Chempedak and Marang. *Tropical Fruits, 2,* 25–51. doi:10.1079/9781845937898.0025

Philippine Information Agency. (2019). BAR intensifies support to jackfruit R&D. Retrieved September 8, 2020, from https://pia.gov.ph/news/articles/1023851

Popp, J., Pető, K., & Nagy, J. (2012). Pesticide Productivity and Food Security. A Review. *Agronomy for Sustainable Development, 33*(1), 243–255. doi:10.1007/s13593-012-0105-x

Rojas-Ruiz, J. (2012). *Quinoa: Economic Growth Hindering Economic Development?* Washington: The Council on Hemispheric Affairs. Retrieved from ProQuest One Academic; Social Science Premium Collection Retrieved from https://link.ezproxy.neu.edu/login?url=https://www-proquest-com.ezproxy.neu.edu/docview/1284595594?accountid=12826

Röös, E., Ekelund, L., & Tjärnemo, H. (2014). Communicating the Environmental Impact of Meat Production: Challenges in the Development of a Swedish Meat Guide. *Journal of Cleaner Production, 73*, 154–164. doi:10.1016/j.jclepro.2013.10.037

Smith, B. (2008). Developing Sustainable Food Supply Chains. *Philosophical Transactions: Biological Sciences, 363*(1492), 849–861. Retrieved September 13, 2020, from http://www.jstor.org/stable/20208471

Soetardji, J. P., Widjaja, C., Djojorahardjo, Y., Soetaredjo, F. E., & Ismadji, S. (2014). Bio-oil from Jackfruit Peel Waste. *Procedia Chemistry, 9*, 158–164. doi:10.1016/j.proche.2014.05.019

Sundarraj, A. A., & Ranganathan, T. V. (2018). Jackfruit Taxonomy and Waste Utilization. *Vegetos- An International Journal of Plant Research, 31*(1), 67. doi:10.5958/2229-4473.2018.00009.5

Swami, S. B., Thakor, N. J., Haldankar, P. M., & Kalse, S. B. (2012). Jackfruit and Its Many Functional Components as Related to Human Health: A Review. *Comprehensive Reviews in Food Science and Food Safety, 11*(6), 565–576. doi:10.1111/j.1541-4337.2012.00210.x

Tabashnik, B. E., Liu, Y., Malvar, T., Heckel, D. G., Masson, L., Ballester, V., … Ferre, J. (1997). Global Variation in the Genetic and Biochemical Basis of Diamondback Moth Resistance to Bacillus Thuringiensis. *Proceedings of the National Academy of Sciences, 94*(24), 12780–12785. doi:10.1073/pnas.94.24.12780

Thillakawardane, T. U. (2009). Market Opportunities for Value-Added Products of Underutilized Fruits and Vegetables [Abstract]. *Acta Horticulturae*, (806), 465–472. doi:10.17660/actahortic.2009.806.58

USDA (2011). USDA Nutrient Data Set for Retail Beef Cuts. Retrieved September 12, 2020, from https://www.ars.usda.gov/ARSUserFiles/80400525/data/beef/retail_beef_cuts02.pdf

Van den Berg, H., & Jiggins, J. (2007). Investing in Farmers—The Impacts of Farmer Field Schools in Relation to Integrated Pest Management. *World Development, 35*(4), 663–686. doi:10.1016/j.worlddev.2006.05.004

part three

The global view

chapter eight

Transparency in European Union food law

Sam Jennings

Contents

EDITORS' NOTE: FOOD SYSTEM TRANSPARENCY IN THE UNITED STATES AS COMPARED TO THE EUROPEAN UNION

The common European market is often compared to that of the United States, but many argue that the commonalities are merely superficial. What the European Union (EU) and the United States have in common is a free movement of goods across relatively large geographic regions, streamlined oversight by European Food Safety Authority (EFSA), on the one hand, and the Food and Drug Administration/

United States Department of Agriculture (FDA/USDA), on the other hand, a common currency, uniformity in food labeling, and several harmonized food safety regulations. This nearly exhausts the commonalities when one looks deeper into the two systems.

Distinctions between the EU and the U.S. food regulatory systems are, for instance, that the United States consists of one nation, the EU of many, and that the precautionary principle governs all of EU food regulation, while the United States generally does not. Zooming in further, however, shows that these relationships can be qualified and even switched in some instances. For cottage foods, as an example, the precautionary principle and biotech approaches in the United States and the EU are reversed.[a] Specifically,

> preparing Cottage Foods is more readily permitted in the EU than in the United States This is the opposite of typical regulatory schemes, where the EU usually follows a precautionary approach and the United States does not. On the one hand, the EU's preemptive regulation demands little state-level legislation to allow Cottage Food operators to function in local economies. On the other hand, the United States defers to the states to enact individual and, as this article shows, varying Cottage Food regulation that is very state-specific. This flipped EU-U.S. regulatory relationship is unusual in the context of food and environmental law. This article questions whether the U.S. restrictions on Cottage Food operators are warranted consumer protection measures, or whether large-scale producers are keeping their grassroots competitors at bay through such regulations.[b]

The free movement of goods is also warranted by North American Free Trade Agreement (NAFTA), and now United States–Mexico–Canada Agreement (USMCA), to a great extent, while certain goods even within the EU are not harmonized, such as those protected by geographical indications. This comparison could be a fascinating study and has been the subject many books written, but it is beyond the scope of the present one. The question and the short exercise embarking upon just the beginning strokes of this comparison,

---

[a] Gabriela Steier, Alberto G. Cianci, Cottage Foods as Radical Acts of Food Sovereignty: New Perspectives on US and EU Food Regulation, *Vermont Law Journal* (Forthcoming 2020)
[b] Id.

however, provide a logical path toward the contrasting of the U.S. and the EU transparency aspects of food regulation.

In Chapter 8, Sam Jennings shares her vast expertise in the European food system. She explains what makes up food system transparency and how the different factors come together. Similar to Chapter 2 by experts Petrenko and Tutelyan, Jennings zooms in on the components of risk assessment, risk management, and risk communication. Where Chapter 2 explains these terms in the context of the Codex Alimentarius, Food and Agriculture Organization (FAO), World Health Organization (WHO), and World Trade Organization (WTO), Jennings elegantly explains them in the following chapter with emphasis on EFSA.

Introduction

This chapter introduces the risk assessment, risk management, and risk communication in the European Union (EU), including the newest updates. First, the chapter provides the regulatory set up of the EU, then explains which bodies have oversight, and provides some examples. Next, the chapter lays out the difficulty that lies in finding the right balance between transparency of the process, highlighted as a consumer need by the regulatory fitness and performance (REFIT) process, while maintaining essential confidentiality of commercial information. At the heart of the chapter are the transparency requirements and their applicability in the EU, thoroughly detailed and contextualized with cases and examples. The EU's new transparency regulations and amendments of existing EU regulation introduce new sections laying down the objectives and general principles of risk communication, which are summarized and explained in this chapter.

Setting the foundation for food risk assessments in the EU

The EU is an economic and political union of twenty-seven Member States comprised from European countries. These countries have met the membership criteria and signed up to an agreement to belong to the EU. There were twenty-eight Member States, but the United Kingdom (UK) withdrew from its membership agreement and left the EU on January 31, 2020. At the time of writing, the UK continues to follow EU law during a one-year transition period, but will have officially exited the EU on January 31, 2021, in a process popularly referred to as "Brexit."

The EU allows for a Single Market based on four freedoms: the free movement of goods, capital, services, and persons between its Member States. There are three further countries that also benefit from this Single Market, as they form part of the larger European Economic Area (EEA). The EEA is an international agreement that comprises the twenty-seven EU Member States plus Iceland, Liechtenstein, and Norway.[1] A large part of the EEA agreement necessitates these three additional countries to implement a considerable number of EU laws into their own national legislation. However, they do not have a great deal of input into the preparation of these laws. It appears to be unlikely that the UK will become part of the EEA.

To enable the free movement of food goods throughout the Single Market, there are wide-ranging EU laws that cover food products from "farm to fork." These food laws have been developed, refined, and increased since the concept of a Single European Market first began to become a possibility with the signing of the Treaty of Rome by Belgium, France, Germany, Italy, Luxembourg, and the Netherlands in 1957.[2] The free movement of goods includes food and feed. It follows that EU food law has as overarching principles, the requirements of Regulation (EC) 178/2002 of the European Parliament and of the Council, which lay down the general principles and requirements of food law, establishing the European Food Safety Authority (EFSA) and laying down procedures in matters of food safety.[3] This Regulation is known colloquially as the Regulation on General Food Law and it sets the foundation of risk analysis on which much of EU food law is based.

Risk analysis in EU food law

Defining the components of risk analysis

In the context of EU food law, risk analysis is comprised of three components: risk assessment, risk management, and risk communication, with the aim of providing a "systematic methodology for the determination of effective, proportionate, and targeted measures or other actions to protect health".[3]

Risk assessment is defined by Regulation (EC) 178/2002 as "a scientifically based process consisting of four steps: hazard identification, hazard characterization, exposure assessment and risk characterization". The EU wanted to ensure that risk assessments could be undertaken in an independent, objective, and transparent manner, in order to encourage public confidence in the scientific basis for food law. In order to enable such independent assessments, the Regulation on General Food Law established the EFSA as the governing body ultimately responsible for all European-wide risk assessments of food in the EU. Member States of the EU may have their own national risk assessment bodies for undertaking assessments for domestic areas of concern. However, the potential risk or intended national actions against this perceived risk may have a wider impact since the EU relies on the EFSA to ultimately develop risk assessments.

When a national body determines a potential risk to their population in relation to a currently harmonized food substance, the Member State should alert the European Commission of this concern. The issue would generally be discussed between the European Commission and all Member States, and if considered necessary, a decision is taken to refer the concern to EFSA for a full safety assessment in relation to the EU population. EFSA's opinion is then taken into account by the European Commission and Member States when setting the risk management measures for the EU as a whole. This use of EFSA as the ultimate risk assessor helps maintain harmonization across the EU.

If individual Member State authorities decide to take national risk management measures independently, based on their own risk assessment body's opinion and without waiting for an EFSA opinion and associated harmonized risk management measures, this can have a detrimental impact on trade with that country both by other EU Member States and by countries outside of the EU. Such a situation is currently faced with the authorized food additive (excipient) color E 171 titanium dioxide. EFSA has produced risk assessments in relation to this color, which to date have indicated no need for risk management measures. However, France brought some more recent studies to the European Commission's attention and EFSA has been asked to undertake an updated risk assessment. In the meantime, France has proceeded to prohibit the use of E 171 titanium dioxide in all food products, whereas all the other Member States are waiting on the outcome of EFSA's revised risk assessment. As E 171 titanium dioxide is an authorized food additive that appears on a harmonized EU positive list of technological food additives, this has had major implications on the import into France of food products containing this substance.

Risk management is defined by the Regulation on General Food Law as "the process, distinct from risk assessment, of weighing policy alternatives in consultation with interested parties, considering risk assessment and other legitimate factors, and, if need be, selecting appropriate prevention and control options." Risk management in the EU is undertaken by the EU Commission in conjunction with the EU Member States, with input of comments and data from stakeholders such as industry and consumer groups. Most areas of food and feed policy have dedicated EU Commission and Member State working groups, which are attended by representatives of the relevant national policy area from Member States and by the relevant personnel from the EU Commission. These working groups discuss the technical and health aspects of areas of potential concern, determine whether policy changes should be proposed for implementation, and draft, revise, and put forward for adoption any proposed

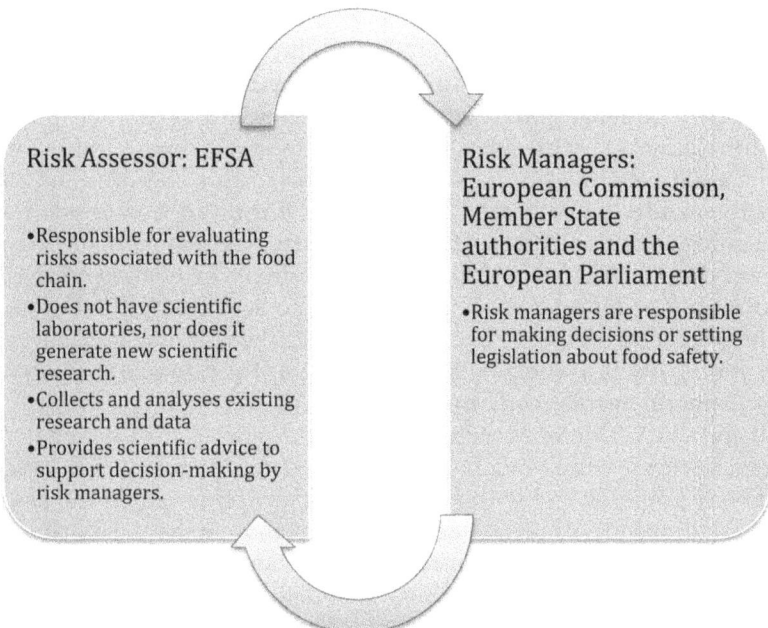

Figure 8.1 Risk assessment and risk management in EU food safety. The two sides visualize the connection between risk management and risk assessment and the responsible bodies in the EU. (This graphic is adapted from Risk Assessment vs Risk Management: What's the difference? EFSA, https://ec.europa.eu/food/sites/food/files/safety/docs/efsa_infographic_roles-of-risk-assessors_en.pdf [last accessed June 28, 2020].)

measures. Although the risk management measures are discussed in depth by the EU Commission and the EU Member States, it is the EU Commission that has ultimate responsibility for any measures that may be put forward.

According to EFSA's own explanation of its scope of oversight, risk assessment is about "[p]roviding scientific advice on food-related risks to support decision-making."[c] However, risk assessment under EFSA explicitly does not include the following:

- Policy making on food safety
- Setting or enforcing legislation

[c] *Risk Assessment vs Risk Management: What's the difference? EFSA*, https://ec.europa.eu/food/sites/food/files/safety/docs/efsa_infographic_roles-of-risk-assessors_en.pdf *(last accessed June 28, 2020)*

- Product approvals and authorizations, recalls, and withdrawals
- Food labeling
- Food quality
- Trade issues, import/export controls, traceability
- Investigation of food fraud

Risk communication is defined as "the interactive exchange of information and opinions throughout the risk analysis process as regards hazards and risks, risk-related factors and risk perceptions, among risk assessors, risk managers, consumers, feed and food businesses, the academic community and other interested parties, including the explanation of risk assessment findings and the basis of risk management decisions." It is to be undertaken in different ways applicable to their roles by EFSA, by the EU Commission, and by individual EU Member States, with additional input as relevant from industry, academia, and consumer groups.

Protection of human health and life

Risk analysis supports the implementation of one of the key objectives of EU food law, namely maintaining a high level of protection of human health and life. It is such protection of human health and life that forms the basis of most current EU food legislation.

Within the EU, any food placed on the market must be safe, and it is the responsibility of food business operators at all stages of the food chain to ensure the safety of the food products under their control. As part of that responsibility, food business operators are obliged to check that the food product and/or its components are compliant with all aspects of food law. As a nonexhaustive list, this includes ensuring that the food and its components are authorized for use in food in the EU (via laws on novel foods,[4] vitamin and minerals and their sources,[5] products of animal origin,[6] food additives (excipients),[7] food flavorings,[8] food enzymes used for technological purposes);[9] that the food and its components are fully compliant with any relevant compositional criteria, purity criteria, maximum levels of chemical contaminants[10] and microbiological contaminants,[11] maximum residue levels of pesticides[12] and/or veterinary products,[13] permitted extraction solvents and their residues,[14] and controls on genetic modification,[15,16] irradiation[17,18] and radioactive contamination;[19] in addition to traceability,[3] allergen control and full labeling requirements.[20] Figure 8.2 describes the relationship between risk assessment and risk management principles in the European food supply chain.

The laws setting out the positive and negative lists, various criteria, and controls have the safety of the consumer as their driving principle, and they have been created and are amended or replaced following appropriate risk analysis procedures.

Risk Assessment	Risk Management
EFSA carries out risk assessment on safety of certain neonicotinoids for bees	Risk managers suspend use of certain neonicotinoids in EU
EFSA evaluates safety of every GMO on a case-by-case basis	Risk managers decide whether or not to authorise each GMO
EFSA collects and analyses data from Member States on prevalence of Salmonella in poultry holdings and assesses risk for human health	Risk managers set reduction targets for Salmonella in laying hens in the EU

Figure 8.2 Examples of risk assessment and risk management in the EU. This table contrasts the risk assessment and risk management sides of three examples, neonicotinoids, genetically modified organisms (GMOs), and *Salmonella*. (It is adapted from Risk Assessment vs Risk Management: What's the difference? EFSA, https://ec.europa.eu/food/sites/food/files/safety/docs/efsa_infographic_roles-of-risk-assessors_en.pdf [last accessed June 28, 2020].)

Transparency in EU risk analysis

Past, present, and future

Chapter II Section 2 of Regulation (EC) 178/2002 focuses on the principles of transparency. It sets out requirements for open and transparent public consultation, directly or through representative bodies, during the preparation, evaluation, and revision of EU food law, except where the urgency of the matter does not allow it.

There are also requirements laid down for EU authorities to make public certain relevant detailed information where there are reasonable grounds for them to suspect that a food or feed may present a risk for human or animal health. The public information should include a description of the particular risk to health, the specific food or feed (or type of food or feed) involved, the risk that the food or feed may present, and the measures that have been taken or are about to be taken to prevent, reduce, or eliminate that risk. Such announcements apply to all food products and can often be seen on a national authority's website.[21,22] They are usually replicated in standard formats as required by a particular country in the media, retail outlets, and/or other appropriate locations. One example of this was the recall of "Eat Natural Brazil & Sultana with Peanuts and Almonds Bar 50G" because of the possible presence of Salmonella. The recall notice was published on the company's own website in the language of each region in which it is marketed on August 22, 2020.[23] The Dutch Food and Consumer Product Safety Authority also published the alert on August 22, 2020[24] while the Food Safety Authority of Ireland published it on August 25, 2020.[25] The major supermarkets in both countries published recall notices to their customers around these dates (for example, Tesco made the alert public on August 25, 2020[26]) and the alerts appeared in regional and national newspapers and online news sources.

In addition to the principles of transparency as set out in Chapter II Section 2, transparency requirements of EFSA and how it conducts its assessments are addressed in Chapter III of the Regulation on General Food Law. The transparency of EFSA's procedures is considered an essential condition of its position as a point of reference, and any scientific studies considered necessary for the performance of EFSA's mission should be commissioned in an open and transparent fashion.

Section 4 of Chapter III deals specifically with EFSA's independence, transparency, confidentiality, and communication, with Article 38 of the Chapter laying down the requirement for EFSA to undertake its activities with a high level of transparency. In particular, EFSA is required to make the following information public as soon as possible after the event:

- agendas, minutes, and opinions of the scientific bodies;
- the information on which its opinions are based, without prejudice to the requirements of confidentiality and access to documents (as also laid down in Chapter III);
- the annual declarations of interest made by members of the various bodies of EFSA, and the declarations of interest made in relation to items on the agendas of meetings;
- the results of its scientific studies;
- the annual report of its activities;
- requests from the European Parliament, the EU Commission, or an EU Member State for scientific opinions which have been refused or modified, and the justifications for the refusal or modification.

In addition to the above, EFSA's Management Board is required to hold its meetings in public, unless a decision is made otherwise for specific administrative points of its agenda, and the Board can also authorize consumer representatives or other interested parties to observe the proceedings of some of EFSA's activities.

2018 REFIT evaluation

In 2018, the EU Commission completed its regulatory fitness and performance (REFIT) evaluation of Regulation (EC) 178/2002 on general food law. The REFIT program has been in operation for a number of years, and looks at EU laws across all sectors and areas. This evaluation aims (i) "to ensure that EU legislation delivers results for citizens and businesses effectively, efficiently and at minimum cost"; and (ii) "to keep EU law simple, remove unnecessary burdens and adapt existing legislation without

compromising on policy objectives".[27] In simple terms, the REFIT evaluation of the Regulation on General Food Law assessed whether it was still
fit for purpose in the current time.

There were a number of positive findings from the evaluation, including the fact that since its application into law, the Regulation on General
Food Law had led to better transparency of the EU decision-making cycle.
However, among the negative findings of the evaluation, it was found
that lack of transparency of risk analysis was considered an important
issue to improve upon consumer perception for EFSA's independent scientific work.

At the risk assessment stage, the difficulty lies in finding the right
balance between transparency of the process while maintaining essential confidentiality of commercial information. As stated above, the EU
positive and negative lists of substances, various criteria, and controls, are
created, amended, or replaced following appropriate risk analysis procedures. The assessments required as part of this risk analysis frequently
necessitate the submission by commercial businesses of detailed data pertaining to their products. When it comes to amending the positive lists of
substances, this submission of data is compulsory if the company's dossier is to be considered valid by the EU.

The original general food law Regulation allowed for a company
to request aspects of the dossier be kept confidential (within certain
limitations), with a nonconfidential summary being made available to the
general public. However, this retention of confidentiality, while ensuring
commercial business confidence in the risk assessment process, led other
areas of the EU population to consider there to be a lack of transparency
and independence in EFSA's work, resulting in reduced credibility of its
scientific output. The REFIT evaluation thus identified a need to address
this issue in order to protect the reputation of EFSA's work.

In addition to the concerns relating to the risk assessment process, the
REFIT evaluation highlighted a lack of effectiveness of risk communication throughout the risk analysis process, which had resulted in reduced
consumer trust and a negative impact on the acceptability of risk management decisions.

The new transparency regulation

The REFIT evaluation of Regulation (EC) 178/2002 on general food law
resulted in the recognition of the need for improved transparency and
also sustainability of the legal processes. Thus, in 2019, Regulation (EU)
2019/1381 of the European Parliament and of the Council on the transparency and sustainability of the EU risk assessment in the food chain
was published.[28] This new Transparency Regulation amends a number
of areas of Regulation (EC) 178/2002 on general food law and also makes
appropriate amendments to a number of other pieces of EU food law, in

order to guarantee consistent and transparent risk communication during the entire risk analysis process.

Amendments to Regulation (EC) 178/2002 on general food law include the insertion of new sections laying down the objectives and general principles of risk communication. Under the new Transparency Regulation, the objectives of risk communication include, among others, the need to raise awareness and understanding of the specific issues being addressed; to increase public understanding of the risk analysis procedure, including the reasons behind the risk management decisions; and to ensure consistency, transparency, and clarity when formulating risk management recommendations and decisions. There is a requirement for the European Commission to adopt a general plan for how these risk communication objectives will be met, and for Implementing Acts to be created in order to fulfil them. Their aim is to have an integrated risk communication framework that will be followed by both the risk assessors and the risk managers in a coherent and systematic manner, both at Union level and at national level within the individual Member States.

Changes to EFSA's structure and procedures
The Transparency Regulation makes certain amendments to the composition and operation of EFSA's Management Board and Scientific Committee, and includes a new requirement for EFSA to provide presubmission advice on the rules applicable to, and the content required for, the particular type of submission, if requested by an applicant.

In addition, EFSA is tasked with maintaining a database of studies that are to be carried out to support an application or notification, and applicants and laboratories must notify EFSA of the relevant details when the studies are being commissioned. These requirements will also apply to countries outside of the EU, where current bilateral or other agreements exist.

An application or notification will not be considered valid if it is supported by studies that have not been previously notified unless the applicant or notifier provides a valid justification for the non-notification of such studies. Similarly, if previously notified studies are not included in the application or notification, that application or notification will also be considered invalid, unless the applicant or notifier is able to provide a valid justification for the noninclusion of the studies. As an additional point, if there are serious controversies or conflicting results, the European Commission has the ability to request EFSA to commission scientific studies with the aim of verifying the evidence that has been used in its risk assessment process. It can be seen that these new requirements have the goal of ensuring that the scientific assessments and resultant opinions are based on an entirety of verified data, with no possibility of unexplained omissions.

Confidential data

Of considerable concern to industry during the drafting of Regulation (EU) 2019/1381, and currently in terms of its implementation, are the requirements surrounding confidential data. Once EFSA receives an application or notification, it has to publish without delay the scientific data, studies, and other information submitted in support of the application or notification, including any supplementary information supplied by the applicants.

The impact on industry of this requirement has been offset slightly by the provision for EFSA not to make public any information for which confidential treatment has been requested. However, there are restrictions surrounding the type of data for which confidentiality may be requested, and also the need for EFSA to concur that the data merits confidential status. An applicant can request confidentiality only for the following types of data:

- the manufacturing or production process, including its method and innovative aspects, plus other technical and industrial specifications that are intrinsic to that process or method, with the exclusion of any information that is relevant to EFSA's safety assessment;
- commercial links between a producer or importer and the applicant or authorization holder (in the case of applications for authorization renewal);
- commercial information revealing sourcing, market shares, or business strategy of the applicant; and
- the quantitative composition of the subject matter of the request, except for data of relevance to the safety assessment.

When the application or notification is submitted, the applicant has to provide a verifiable justification of why certain of the above areas must be kept confidential, demonstrating how the making public of the data would significantly harm its interests. In such cases, the applicant must also provide a nonconfidential version of the application, which will be made public by EFSA as soon as it has been confirmed as valid or admissible.

The justification supporting the need for confidentiality will be assessed by EFSA. If the justification for confidentiality of certain data is accepted, the safety assessment can proceed. However, if EFSA does not agree with the request for confidentiality of certain information, it has to inform the applicant, who then has two weeks in which to either withdraw the application or submit a confirmatory application asking EFSA to reconsider its decision. If neither of these actions are taken by the applicant, EFSA will publish the data and information for which the confidentiality request has not been accepted as justified.

If the applicant submits a confirmatory application asking EFSA to reconsider its decision, EFSA then has three weeks in which to review the request and adopt a reasoned decision. It will then notify the applicant of its decision and include within that notification the information that the applicant can take a case forward to the Court of Justice of the EU against EFSA. Meanwhile, the additional data and information for which the confidentiality request was not accepted will be published no less than two weeks after EFSA notified the applicant of its reasoned decision.

Other revisions to the general food law regulation

Rules are additionally added into Regulation (EC) 178/2002 relating to the protection of certain personal data relating to applications or notification; the implementation of standard data formats for use by applicant; the use by EFSA of auditable and secure information systems; the ability for the European Commission to review, every five years, EFSA's performance in relation to its objectives, mandate, tasks, procedures, and location, with the outcome of the review being made public; and for relevant experts from the European Commission to undertake "fact-finding missions" in EU Member States and in countries outside of the EU (as far as is permitted by relevant agreements and arrangements), confirming the performance of laboratories and other testing facilities involved in undertaking studies and providing data used in submissions to EFSA.

Revisions to other legislation

In addition to the key revisions to Regulation (EC) 178/2002 on general food law, Regulation (EU) 2019/1382 makes amendments to the following legislative acts and, in some cases, to their related Implementing Regulations:

- Regulation (EC) No 1829/2003 on genetically modified food and feed;[15]
- Regulation (EC) No 1831/2003 on the traceability and labeling of genetically modified organisms and the traceability of food and feed products produced from genetically modified organisms;[16]
- Regulation (EC) No 2065/2003 on smoke flavorings used or intended for use in or on foods;[29]
- Regulation (EC) No 1935/2004 on materials and articles intended to come into contact with food;[30]
- Regulation (EC) No 1331/2008 establishing a common authorization procedure for food additives, food enzymes, and food flavorings;[31]
- Regulation (EC) No 1107/2009 on the placing of plant protection products on the market;[32]

- Regulation (EU) 2015/2283 on novel foods;[4]
- Directive 2001/18/EC on the deliberate release into the environment of genetically modified organisms.[33]

The amendments to the above pieces of legislation add into the texts the additional requirements for transparency and control of confidential data as laid down in the revised general food law Regulation. For certain of the above, where innovative methods or starting materials may be key to the submission, clarity is provided regarding other data for which confidential status might be accepted.

For example, for Regulation (EC) 1829/2003 on genetically modified (GM) food and feed, a submission for assessment of a new GM food or feed product may include a request for confidential treatment of the areas highlighted in the revised general food law Regulation and also the following:

- DNA sequence information, except for sequences used for the purpose of detection, identification, and quantification of the transformation event; and
- breeding patterns and strategies.

Similarly, for Regulation (EU) 2015/2283 on novel foods (which lays down the requirements for a safety assessment of a food or ingredient that was not on the EU market to a significant degree prior to May 1997), confidential status may additionally be requested for:

- information provided in detailed descriptions of starting substances and starting preparations, and on how they are used to manufacture the novel food;
- detailed information on the nature and composition of the specific foods or food categories in which the applicant intends to use the novel food, except for information which is relevant to the assessment of safety;
- detailed analytical information on the variability and stability of individual production batches, except for information which is relevant to the assessment of safety.

Comparable allowances are made in relation to confidential data for the other six amended legislative acts.

Implementation phase

The measures laid down in the Transparency Regulation (EU) 2019/1381 will not apply until March 27, 2021, with the exception of the changes to EFSA's structure, which will apply from July 1, 2022. This delay in application gives the European Commission and Member States time to develop, discuss, and adopt the general plan for meeting the risk communication objectives and the Implementing Acts needed to fulfil them – the "implementation phase."

During this period, the European Commission and EFSA are working closely together to ensure that the Transparency Regulation is implemented appropriately and transparently, with the engagement of all stakeholders. To achieve this aim, the Commission and EFSA jointly agreed and aligned on a stakeholder engagement framework,[34] which operates to the principles of independence, transparency, openness, inclusiveness, flexibility, and nonduplication.

To ensure independence, transparency, and openness, the only interaction by the European Commission and EFSA with stakeholders is via dedicated meetings where updates on progress are summarized. Stakeholders can continue to submit comments outside of these forums, but individual responses are not given. Instead, the comments will be posted on the European Commission or EFSA's website,[35,36] depending on which was the recipient. If the stakeholder gives consent, the entire document is made publicly available. If the stakeholder does not wish their comments to be made public, the following will be published on the website: the sender's name and affiliation, the title of the document and a statement to the effect that the sender did not give their consent for publication of the document on the website. The Commission or EFSA respond to questions, positions, feedback, or comments only during the dedicated stakeholder meetings, where all parties can hear the discussion. Information coming from the European Commission or EFSA has to be provided promptly and with clarity, and their websites include the agendas and reports or notes from various meetings relevant to this implementation phase.

The principle of inclusiveness is followed by treating all stakeholders equally during the dedicated meetings. Where these meetings are for general updates, all categories of stakeholders are invited. Should expert input from a particular category of stakeholders be required, for example, in EFSA technical meetings, other stakeholder categories are invited as observers.

Flexibility is to be maintained by including public consultations when needed alongside the other meetings, whether physical or held online, while where possible, duplication is avoided between other previously existing stakeholder engagement meetings and those meetings dedicated to the work on implementation of the Transparency Regulation.

At the time of writing, the European Commission is in the process of developing the Implementing Acts required for fulfilling the objectives of the Transparency Regulation. Once prepared, these will be put out for public consultation via the Commission's feedback mechanism, which provides a four-week window for comments. Relevant guidelines are being revised where necessary, and the practicalities of the Commission's "fact-finding missions" are being discussed.

In July 2020, three draft Implementing Acts were published for a four-week public consultation via the European Commission's feedback mechanism. Two of these drafts made amendments to novel foods legislative acts: Commission Implementing Regulation (EU) 2017/2469 of 20 December 2017 laying down administrative and scientific requirements for applications referred to in Article 10 of Regulation (EU) 2015/2283 of the European Parliament and of the Council on novel foods;[37] and Commission Implementing Regulation (EU) 2017/2468 of 20 December 2017 laying down administrative and scientific requirements concerning traditional foods from third countries in accordance with Regulation (EU) 2015/2283 of the European Parliament and of the Council on novel foods.[38] The third draft made amendments to Commission Regulation (EU) No 234/2011 of 10 March 2011 implementing Regulation (EC) No 1331/2008 of the European Parliament and of the Council establishing a common authorization procedure for food additives, food enzymes, and food flavorings.[39] The proposed revisions to the original legislative acts were as had been expected, following the requirements of the Transparency Regulation.

EFSA has published an amendment to its 2003 Decision concerning access to documents.[40] The amended Decision sets out the revised process for the handling of applications for access to documents held by EFSA, including the procedures highlighting the steps required at each stage for replying to applications for access to these documents; the conditions for applying exceptions to disclosure; and the means by which all legal rights and obligations are reconciled during the entire process. EFSA has additionally established two technical groups, one that focusses on the use of the European Chemicals Agency (ECHA)'s IUCLID cloud for pesticide dossiers, and the other that is working on the Notification of Studies Database. In addition to representatives from EFSA, the Technical Groups are comprised of relevant stakeholder representatives of EU agencies, Member States, the European Commission, and observers. All relevant documents can be accessed on the EFSA Transparency Regulation implementation and stakeholder engagement webpage.[41]

Conclusion

It can be seen that the European Commission is intent on increasing transparency across the risk analysis process, thereby improving the EU population's viewpoint of the scientific opinions and regulatory decisions that come out of the process. There are aspects of this increased transparency that are causing concerns within industry, in particular, the processes surrounding confidentiality of data, and the impact publication of such data might have on a business should EFSA not agree that the request for confidentiality be merited. Interested stakeholders are therefore following the Implementation Phase very closely; when a draft Implementing Act is published for consultation, there will be only four weeks in which to review, digest, and comment on the text, so delays in viewing the consultation will need to be avoided.

In general, the European Commission has over the years tried to be transparent in its work in other areas of food law, working with stakeholders during the development phase of new or amending legislative acts; requesting, accepting, and discussing evidence that may impact the regulatory decisions; sharing early draft texts; and being open to suggested revisions to reduce any detrimental impacts on employment, internal and international trade, and consumer choice. There have in recent years been a few situations where the European Commission has been far less transparent, keeping all development work internal within its relevant departments, and providing no indication to stakeholders or even Member States as to its proposed regulatory action until publication is made via the feedback mechanism. Fortunately, as at the time of writing, these situations are still a rarity, as most departments in the European Commission recognize the importance of early discussion and interaction with Member States, relevant stakeholders, and other interested parties.

It is as yet unclear as to how or whether the UK will implement the Transparency Regulation in its national law following the end of the Brexit transition period on December 31, 2020. The Transparency Regulation itself will be carried across into UK law, as "retained EU law." However, as the measures in this Regulation do not apply until March 2021, and because the UK has not been involved in the discussions relating to how the measures will be implemented in the EU, it is unknown whether the same measures will eventually apply to the UK or whether the decision will be taken not to apply the Transparency Regulation within the UK. Similarly, the EEA countries may choose not to apply the measures to applications made nationally.

Regardless of whether the transparency measures are applied nationally in the UK or the EEA countries, they will apply to all relevant applications made in relation to food intended for the EU market. For example, a company based in the United States, making an application for a novel food authorization in the EU, will face the same transparency requirements as those of an EU-based company making a similar application. Therefore, companies based outside of the EU, who are considering submitting applications under any of the affected legislative food areas, will need to be prepared for the additional application requirements and the impacts these might have on their submission and, hence, the potential for future trade with the EU.

Consumer safety forms the basis of the majority of European food law, but consumer confidence in the law is now recognized as also being of importance. By increasing transparency in its risk analysis processes, public opinion of the law can improve in the future.

References

1. The European Free Trade Association. The Basic Features of the EEA Agreement. https://www.efta.int/eea/eea-agreement/eea-basic-features (accessed 29 May 2020)
2. Meulen, B. (2013). The Structure of European Food Law. Laws. 2. 69–98.
3. Regulation (EC) No 178/2002 of the European Parliament and of the Council of 28 January 2002 laying down the general principles and requirements of food law, establishing the European Food Safety Authority and laying down procedures in matters of food safety (OJ L 31, 1.2.2002, p. 1)
4. Regulation (EU) 2015/2283 of the European Parliament and of the Council of 25 November 2015 on novel foods, amending Regulation (EU) No 1169/2011 of the European Parliament and of the Council and repealing Regulation (EC) No 258/97 of the European Parliament and of the Council and Commission Regulation (EC) No 1852/2001 OJ L 327, 11.12.2015, p. 1–22
5. Regulation (EC) No 1925/2006 of the European Parliament and of the Council of 20 December 2006 on the addition of vitamins and minerals and of certain other substances to foods. OJ L 404, 30.12.2006, p. 26–38
6. Regulation (EC) No 853/2004 of the European Parliament and of the Council of 29 April 2004 laying down specific hygiene rules for food of animal origin. OJ L 139, 30.4.2004, p. 55–205
7. Regulation (EC) No 1333/2008 of the European Parliament and of the Council of 16 December 2008 on food additives. OJ L 354, 31.12.2008, p. 16–33
8. Regulation (EC) No 1334/2008 of the European Parliament and of the Council of 16 December 2008 on flavourings and certain food ingredients with flavouring properties for use in and on foods and amending Council Regulation (EEC) No 1601/91, Regulations (EC) No 2232/96 and (EC) No 110/2008 and Directive 2000/13/EC. OJ L 354, 31.12.2008, p. 34–50
9. Regulation (EC) No 1332/2008 of the European Parliament and of the Council of 16 December 2008 on food enzymes and amending Council Directive 83/417/EEC, Council Regulation (EC) No 1493/1999, Directive 2000/13/EC, Council Directive 2001/112/EC and Regulation (EC) No 258/97. OJ L 354, 31.12.2008, p. 7–15
10. Commission Regulation (EC) No 1881/2006 of 19 December 2006 setting maximum levels for certain contaminants in foodstuffs. OJ L 364, 20.12.2006, p. 5–24
11. Commission Regulation (EC) No 2073/2005 of 15 November 2005 on microbiological criteria for foodstuffs. OJ L 338, 22.12.2005, p. 1–26
12. Regulation (EC) No 396/2005 of the European Parliament and of the Council of 23 February 2005 on maximum residue levels of pesticides in or on food and feed of plant and animal origin and amending Council Directive 91/414/EEC. OJ L 70, 16.3.2005, p. 1–16

13. Regulation (EC) No 470/2009 of the European Parliament and of the Council of 6 May 2009 laying down Community procedures for the establishment of residue limits of pharmacologically active substances in foodstuffs of animal origin, repealing Council Regulation (EEC) No 2377/90 and amending Directive 2001/82/EC of the European Parliament and of the Council and Regulation (EC) No 726/2004 of the European Parliament and of the Council. OJ L 152, 16.6.2009, p. 11–22

14. Directive 2009/32/EC of the European Parliament and of the Council of 23 April 2009 on the approximation of the laws of the Member States on extraction solvents used in the production of foodstuffs and food ingredients (Recast). OJ L 141, 6.6.2009, p. 3–11

15. Regulation (EC) No 1829/2003 of the European Parliament and of the Council of 22 September 2003 on genetically modified food and feed. OJ L 268, 18.10.2003, p. 1–23

16. Regulation (EC) No 1830/2003 of the European Parliament and of the Council of 22 September 2003 concerning the traceability and labelling of genetically modified organisms and the traceability of food and feed products produced from genetically modified organisms and amending Directive 2001/18/EC. OJ. L 268, 18.10.2003, p. 24–28

17. Directive 1999/2/EC of the European Parliament and of the Council of 22 February 1999 on the approximation of the laws of the Member States concerning foods and food ingredients treated with ionising radiation OJ L 66, 13.3.1999, p. 16–23

18. Directive 1999/3/EC of the European Parliament and of the Council of 22 February 1999 on the establishment of a Community list of foods and food ingredients treated with ionising radiation. OJ L 66, 13.3.1999, p. 24–25

19. Council Regulation (Euratom) 2016/52 of 15 January 2016 laying down maximum permitted levels of radioactive contamination of food and feed following a nuclear accident or any other case of radiological emergency, and repealing Regulation (Euratom) No 3954/87 and Commission Regulations (Euratom) No 944/89 and (Euratom) No 770/90. OJ L 13, 20.1.2016, p. 2–11

20. Regulation (EU) No 1169/2011 of the European Parliament and of the Council of 25 October 2011 on the provision of food information to consumers, amending Regulations (EC) No 1924/2006 and (EC) No 1925/2006 of the European Parliament and of the Council, and repealing Commission Directive 87/250/EEC, Council Directive 90/496/EEC, Commission Directive 1999/10/EC, Directive 2000/13/EC of the European Parliament and of the Council, Commission Directives 2002/67/EC and 2008/5/EC and Commission Regulation (EC) No 608/2004. OJ L 304, 22.11.2011, p. 18–63

21. Food Safety Authority of Ireland: Food Alerts. https://www.fsai.ie/news_centre/food_alerts.html (accessed 29 May 2020)

22. Spanish Agency for Consumer Affairs, Food Safety and Nutrition: Food Alerts. http://www.aecosan.msssi.gob.es/en/AECOSAN/web/seguridad_alimentaria/subseccion/otras_alertas_alimentarias.htm (accessed 29 May 2020)

23. Eat Natural Recall Notice. https://www.eatnatural.com/news/eat-natural-recalls-batches-of-its-brazil-sultana-with-peanuts-and-almonds-bar-due-to-possible-salmonella-risk/?back=1 (accessed 7 September 2020)

24. Dutch Food and Consumer Product Safety Authority: Eat Natural Recall Notice. https://www.nvwa.nl/documenten/waarschuwingen/2020/08/22/belangrijke-veiligheidswaarschuwing-eat-natural-brazil-sultana-with-peanuts-and-almonds (accessed 7 September 2020)

25. Food Safety Authority of Ireland: Eat Natural Recall Notice. https://www.fsai.ie/news_centre/food_alerts/eat_natural_bars.html (accessed 7 September 2020)
26. Tesco Eat Natural Recall Notice. https://digitalcontent.api.tesco.com/v2/media/homepage/03668130-d07d-45f2-ab3f-1c2c3c9f384a/Product+Recall+-+Eat+Natural+bar.pdf accessed 7 September 2020)
27. European Commission REFIT Programme. https://ec.europa.eu/info/law/law-making-process/evaluating-and-improving-existing-laws/refit-making-eu-law-simpler-and-less-costly_en (accessed 29 May 2020)
28. Regulation (EU) 2019/1381 of the European Parliament and of the Council of 20 June 2019 on the transparency and sustainability of the EU risk assessment in the food chain and amending Regulations (EC) No 178/2002, (EC) No 1829/2003, (EC) No 1831/2003, (EC) No 2065/2003, (EC) No 1935/2004, (EC) No 1331/2008, (EC) No 1107/2009, (EU) 2015/2283 and Directive 2001/18/EC. PE/41/2019/REV/1. OJ L 231, 6.9.2019, p. 1–28
29. Regulation (EC) No 2065/2003 of the European Parliament and of the Council of 10 November 2003 on smoke flavourings used or intended for use in or on foods. OJ L 309, 26.11.2003, p. 1–8
30. Regulation (EC) No 1935/2004 of the European Parliament and of the Council of 27 October 2004 on materials and articles intended to come into contact with food and repealing Directives 80/590/EEC and 89/109/EEC. OJ L 338, 13.11.2004, p. 4–17
31. Regulation (EC) No 1331/2008 of the European Parliament and of the Council of 16 December 2008 establishing a common authorisation procedure for food additives, food enzymes and food flavourings. OJ L 354, 31.12.2008, p. 1–6
32. Regulation (EC) No 1107/2009 of the European Parliament and of the Council of 21 October 2009 concerning the placing of plant protection products on the market and repealing Council Directives 79/117/EEC and 91/414/EEC. OJ L 309, 24.11.2009, p. 1–50
33. Directive 2001/18/EC of the European Parliament and of the Council of 12 March 2001 on the deliberate release into the environment of genetically modified organisms and repealing Council Directive 90/220/EEC - Commission Declaration. OJ L 106, 17.4.2001, p. 1–39
34. Implementation of the Transparency Regulation and Stakeholder Engagement. https://ec.europa.eu/food/sites/food/files/safety/docs/gfl_transparency_reg-implement_interaction-fwk.pdf (accessed 7 September 2020)
35. European Commission webpage: Transparency and Sustainability of the EU Risk Assessment in the Food Chain. https://ec.europa.eu/food/safety/general_food_law/transparency-and-sustainability-eu-risk-assessment-food-chain_en (accessed 29 May 2020)
36. European Food Safety Authority Webpage: Transparency. https://www.efsa.europa.eu/en/about/transparency (accessed 29 May 2020)
37. Commission Implementing Regulation (EU) 2017/2469 of 20 December 2017 laying down administrative and scientific requirements for applications referred to in Article 10 of Regulation (EU) 2015/2283 of the European Parliament and of the Council on novel foods. OJ L 351, 30.12.2017, p. 64–71
38. Commission Implementing Regulation (EU) 2017/2468 of 20 December 2017 laying down administrative and scientific requirements concerning traditional foods from third countries in accordance with Regulation (EU) 2015/2283 of the European Parliament and of the Council on novel foods. OJ L 351, 30.12.2017, p. 55–63

39. Commission Regulation (EU) No 234/2011 of 10 March 2011 implementing Regulation (EC) No 1331/2008 of the European Parliament and of the Council establishing a common authorisation procedure for food additives, food enzymes and food flavourings. OJ L 64, 11.3.2011, p. 15–24

40. Decision of the Management Board laying down practical arrangements for implementing Regulation (EC) No 1049/2001 and Articles 6 and 7 of Regulation (EC) No 1367/2006. EFSA. http://www.efsa.europa.eu/sites/default/files/documents/wp200327-a2.pdf (accessed 7 September 2020)

41. EFSA Transparency Regulation Implementation and Stakeholder Engagement. https://www.efsa.europa.eu/en/stakeholders/transparency-regulation-implementation (accessed 7 September 2020)

Discussion and Research Questions

1. How do risk assessments compare to risk analysis, risk management, and risk communication in terms of EU law?
2. Who has the burden of ensuring safe food in the EU? How is this different in the United States? Consider the scope of oversight of the food risk assessment and management bodies in the EU and the United States in addressing this question.
3. How does the Transparency Regulation impact trade agreements between the EEA and UK? How does this differ between U.S. trading partners?
4. Should confidential business information be made available to the public? What did the U.S. Supreme Court decide regarding confidential commercial or financial information in *FMI v. Argus Leader?* Do you agree with this decision in terms of competitive harm?
5. Considering the ongoing developments of Brexit, what are the most recent updates or changes to food system transparency?

chapter nine

Transparency from information-based regulation: The case of China and the Asian area

Juanjuan Sun

Contents

EDITORS' NOTE: SHORTENED SUPPLY CHAINS FOR SUSTAINABLE FOOD SYSTEM TRANSPARENCY

Chapter 9 brings many of the discussions from Chapter 8 into focus and context. Professor Juanjuan Sun graciously explains how regulation works in shortened supply chains and she illustrates this with an elaborate case study from China. Now more than ever, the world's eyes are toward China and how they leverage data and technology to make decisions.

Shortened supply chains, the focus of Chapter 9, means that transparency requires standard rules that must trusted and accessible. Throughout the world, many suppliers, producers, processors, and retailers use different documentation methods and create complex webs of supply chains. Without innovative implementation of technology, important public data that is normally lost in supply chain activities are either lost, hidden, or not collected at all. Throughout the next decade, a greater focus on ethical information sharing between governments, businesses, and consumers will increasingly become more important focus areas within the food system.

Imagine, for instance, a food product – such as an apple – is starting its journey from a given tree on a particular farm lot. The data from this apple can be tracked from the moment it moves to from the tree, to a crate, and until the moment it's sold. One of many apples, this single product may be loaded onto a truck and a plane and then distributed into processing centers, distribution centers, and retail environments. Myriad points exist where one can predictably lose track of the apple, unless certain data are accurately collected throughout the entire supply chain to help prevent lost or misleading data and unaccounted food.

Emerging technologies such as blockchain, artificial intelligence, machine learning, and Internet of Things (IoT) focus on data collection, data storage, and data analyzation within supply chains and systems. But these systems and technologies do not necessarily "talk" to one another (i.e., interoperability) and are generally prone to human error or biased predictions. Opportunities exist to enhance data integrity and system interoperability within local, regional, and multinational supply chains – all to illuminate transparency at every point from farm to fork.

This chapter is also contributing to the National Research Programme: Study on the Construction of Legal Structure Regulating GM Food in China (No. 18ZDA147).

Introduction

The food supply chain has evolved over time, and two trends are developing simultaneously. On one hand, the food supply chain has extended over localities, regions, and countries. Globalized food supply chains are getting longer, more complex, and are characterized by a large number of stakeholders. On the other hand, the rises of local food movements all over the world have localized food supplies, especially locally produced food of high quality. Also, with the onset of increasingly popular online food shopping, the direct sale introduced by online platforms between primary producers and consumers has also shortened the supply chain from the perspective of the intermediate organizations involved. Either for long supply chains or short supply chains, transparency is an increasingly essential requirement for food business operators. The purpose of such practices may vary. It may be out of the regulatory provision in order to protect the consumers' right to know and informed decision-making. Alternatively, self-regulation to put corporate social responsibility into practice in an effort to promote transparency is a necessary condition for corporate social responsibility.[1] Therefore, food and information flow along the supply chain means that product information should be available for other stakeholders, such as consumers.

Notably, transparency refers to the concept that people can see through some medium to an object on the other side (Figure 9.1). This means that the requirement on the information transparency and freeflow of product information gives for individuals or organizations a chance to see the truth, without trying to hide or shade the meaning, or altering the facts to put things in a better light,[2] which is often needed when negative aspects of large, industrially produced supply chains are in play.

Undoubtedly, only producers know the real status of their products, and are capable to inform buyers of the characteristics of their products that are then sold. However, the so-called market failure in terms of information asymmetry may allow food business operators to hide or exaggerate information of food characteristics, with the purpose of misleading consumers to buy adulterated or misbranded product at a lower cost (i.e., fraud). As a result, either for protecting consumers or promoting fair business, the Chinese state has intervened in food businesses through legal instruments. Thus, information transparency has become one of justifications for state regulation of the food supply – but it remains the food business operators' obligation to provide product information.

To improve compliance by food business operators, Chinese state regulators have a diverse range of tools from traditional command-and-control approaches such as licensing to voluntary ones such as guidelines for best practice. When it comes to transparency, how do these regulatory tools facilitate information flow along the supply chain and how do they present information more accessibly for informed consumer choices?

---

[1] Wim Dubbink, Johan Graafland, and Luc van Liedekerke, CSR, transparency and the role of intermediate organizations. https://www.montesquieu-instituut.nl/9353000/1/j4nvih7l3kb91rw_j9vvj72dlowskug/vjclg3i5c7dc/f=/artikel_johan_graafland.pdf
[2] Oliver, R., *What is transparency?* McGraw-Hill Companies, 2004, p. 3.

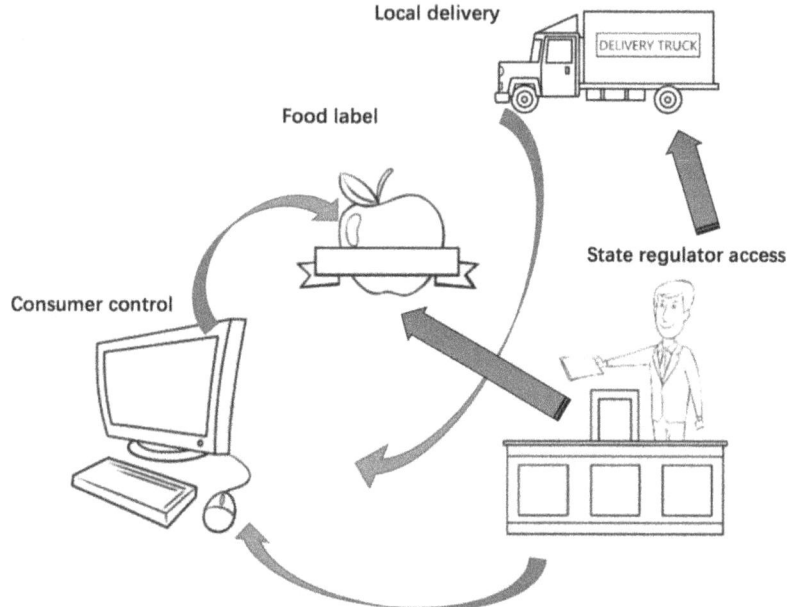

Figure 9.1 Consumer controlled supply chains in online purchases of locally produced foods. This figure illustrates the shortened supply chains when a local food, such as an apple, can be transported directly from farm to table. The delivery truck symbolizes the short supply chain that allows the regulator (person with forms) to directly access both the product label (on the apple) as well as the producer (truck). It is important for consumers to be empowered to ask questions to regulators and producers to understand product characteristics. The curved arrows show the flow of information. The straight lines show mandated data collection requirements within regulations. (Graph by Gabriela Steier.)

Which regulatory tools exist that facilitate this information sharing? To answer these questions, this chapter outlines regulation for the sake of transparency, zooming-in on cases of Asian areas in general, and then the case of China more specifically.

Transparency from the perspective of information regulation

Nowadays, food fraud may be driven by economic gain, and sometimes, it can bring risks or hazards to public health while food business operators put economic gain ahead of safety guarantee. In view of this, state regulation has reinforced food regulation in both food safety protection and food quality promotion. However, food safety attributes and quality attributes are difficult for the consumer to evaluate even after it has been consumed. The nature of food as credence good implies that information that reflects food safety and quality matters. In the case of food safety, this information can refer to residue of pesticide and date of minimum durability. Quality information may address appearance or taste in the form of marketing or geographic indication for a specific area with particular natural and human factors. As mentioned above, it is food business operators that have obligations to provide information for guaranteeing food safety and have interests to promote food quality for comparative advantage. Besides, state regulation carried out by competent authorities also engages in collection, analysis and share of information, and thus contribute to guaranteeing information transparency along the food supply chain.

Food operators and labeling requirements

Given the information asymmetry between sellers and buyers, requiring food business operators to provide information of the product, process within their jurisdiction is a direct way to make sure the information is transparent. In this aspect, food labeling is regarded as the primary means of communication between the different stakeholders such as producer, seller, purchaser, and consumer. As mentioned before, the purpose of such information transparency is to help consumers make informed choices, that is to say, to figure out "what they eat." In this aspect, early legislation for fighting against food fraud in England and United States of America (USA) was aimed to ensure purity and authenticity of food by defining what is so-called misbranded food. Taking the Pure Food and Drug Act of 1906 in the United States as an example, the term "misbranded" shall apply to all drugs, or articles of food, or articles which enter into the composition of food, the package, or label of which shall bear any statement, design, or device regarding such article, or the ingredients or substances contained therein which shall be false or misleading in any particular, and to any food or drug product which is falsely branded as the State, territory, or country in which it is manufactured or produced. For the moment, the Regulation (EU) No. 1169/2011 on food information to consumers in the EU is another valuable case to indicate the significance of information transparency to a high level of consumer protection by improving legibility of mandatory information, presenting certain nutrition information and allergen information and strengthening rules to prevent misleading practice. Meanwhile, as clarified by the EU legislation, harmonization of labeling requirements not only contributes to consumer protection but also fair practices and thus smooth functioning of the market.

Similarly, the update of food labeling legislation in Japan also has multiple purposes. Generally speaking, Japan's Food Labeling Act was enacted in 2013 and implemented on April 1, 2015. Before this legislation, the labeling requirements were scattered in different legislations, like the Act on Standardization of Agricultural and Forestry Products, Food Sanitation Law, and Health Promotion Law. This is why the purposes of the Food Labeling Act is to combine the measures implemented under these different laws. Based on this, integrated labeling requirements are aimed to make an autonomous and rational choice of food, to promote the interests of general consumers, and also to contribute to the protection and promotion of the health of the people, the smooth production and distribution of food, and the promotion of food production in response to consumer demand. To facilitate the law implementation, Japanese Consumer Affairs Agency, as the responsible competent authority, has further specified Food Labeling Standard based on the Food Labeling Act.

As countries may differ on what should be presented in food labels, guidance on food labeling provided by The Codex Alimentarius Commission has harmonized rulemaking at the international level. According to this General Standard for the Labeling of Prepackaged Food, mandatory labeling of packaged foods includes information such as the name of the food, list of ingredients, net contents, name and address of the manufacturer, and country of origin. In addition to the efforts made on the official control, private operators are also engaging in the harmonization of food rules to respond to the consumer's needs, who may be interested or invested in a fair and sustainable food supply chain. To provide this information, private standards and logos for indicating third party certification are mixed to provide food products that can meet a variety of consumer preferences. For instance, the global and increasing demand of palm oil has led to a rapid growth in palm oil production and trade. Meanwhile, such expansion has caused different kinds of problems, such as land conflicts and negative environmental effects. As a response, private governance in the form of the Round Table on Sustainable Palm Oil was initiated, in order to develop a set of principles and criteria for sustainable production of palm oil on the basis of a series of worldwide multiple stakeholder consultations. While the Indonesian government has supported such voluntary private standard, it has also started to develop its own standard in the form of Indonesian Sustainable Palm Oil since it has found that the private standards are insufficient to address the producer responsibility.[3]

Competent authority and information-based official control

Many tools aimed at facilitating official control can be indirectly used to collect and disseminate information and thus improve the information

---

[3] Otto Hospes, Private law making at the round table on sustainable palm oil, in Bernd van der Meulen (eds), *Private Food Law*, Wageningen Academic Publishers, 2011, pp. 187–201.

transparency along the food supply chain. For example, competent authorities are obliged to publish administrative punishment information according to the requirements provided by the legislations on information disclosure. With this information, both food business operators and consumers can know who provides them with food, and they may prefer the food provided by law-abiding operators to reduce their own risks. In view of the limited resources of state regulation, smart regulation at the side of competent authorities has taken advantage of information disclosure to promote the participation of other stakeholders in supervision. A typical case is to make the inspection result available on the business' premise for public reviewing, such as the sanitation grade of a licensed restaurant. Different from the publication of administrative penalty via the internet on sites like the official website, information note for public review tries to provide information in a simple and convenient way. One example is the smiley scheme in Denmark that presents the inspection results of food establishments in the form of a symbol, which is a smiley face ranging from big smile to sad.[4]

To facilitate the public review, such grading systems for food establishments are also popular in Asian countries. Taking Singapore as an example, the National Environment Agency is responsible for licensing food retail businesses, such as restaurants. Based on the unannounced inspection of personnel and food hygiene, and housekeeping of the premises, the National Environment Agency has introduced following grading system to encourage establishments to improve their grade by adopting better practices. The grades standards are given by letter grades as follows, the letter grade A standard for a score of 85 percent or higher, B for a score of 70–84 percent, C for a score of 50–69 percent and D for a score of 40–49 percent. In practice, such a grading system has two shortcomings. For one thing, the assessment of letter grade was only based on a snapshot of hygiene status at the point of an unannounced inspection. For another thing, 99 percent of licenses already have A or B grade and make it more difficult for consumers to be able to distinguish between the better performers from among this pool of licensed businesses. Therefore, a new food hygiene food recognition scheme will replace the old system to award the bronze award, silver award, and gold award if hygiene trace record without major hygiene lapses of two years, five years, and ten years, respectively.[5]

---

[4] Information note, food hygiene information system in selected places, Legislative Council Secretariat, IN19/07-08, available at https://www.legco.gov.hk/yr07-08/english/sec/library/0708in19-e.pdf

[5] New food hygiene award scheme to replace current grading system by 2020, 23 June 2018, available at https://www.gov.sg/news/content/new-food-hygiene-award-scheme

Education and communication

Either for understanding the labeling information or making use of information posted in restaurants, consumers as targeted recipients need to be aware of food safety and be able to interpret the information with professional knowledge. For this reason, state intervention has also given attention to education and information campaigns about food safety regulation. At this point, it is worth mentioning the role of risk communication to promote the interaction between regulators and other stakeholders such as consumers. As a component of well-structured risk analysis, risk communication is the interactive exchange of information and opinions throughout the risk analysis process concerning risk, risk-related factors and risk perceptions, among risk assessors, risk managers, consumers, industry, the academic community, and other interested parties. This includes the explanation of risk assessment findings and the basis of risk management decisions.[6] The significance of risk communication throughout the risk analysis are multifaceted, such as promoting the understanding of the specific issue, the transparency in decision-making, fostering public trust and confidence, etc.[7] Therefore, with the passing of the risk analysis legislation into food laws, risk communication as well as risk assessment have reshaped science-based food safety regulation with stakeholders and public participation.

As far as information dissemination is concerned, the institutionalization of risk communication into food safety regulation has allowed government to be a trusted information provider, in particular to fight against rumors about food safety. Taking South Korea as example, a food safety information center was mandated by an amendment to the Food Sanitation Act, and finally established in 2008. Accordingly, this center should collect, analyze, or provide information concerning food safety in Korea and abroad. After being renamed to the National Food Safety Information Service in 2012, the center has established as one of its missions, to be the most trusted food safety information provider by expanding a demand-based tailored information service.

The case of China

Legally speaking, all above-mentioned regulation has been integrated to Chinese Food Safety Law. In practice, the implementation of these tools in a mixed way as well as other locally promoted tools such as punitive damages and credit system has contributed to improving transparency in

---

[6] Codex Alimentarius Commission, *Procedural manual*, Twenty-first edition, 2013, p. 114.
[7] FAO/WHO, *The application of risk communication to food standards and safety matters*, Report of a Joint FAO/WHO expert consultation, Rome, February 2–6, 1998, p. 4.

the food supply chain. Comparatively, either food operators or competent authorities have taken advantage of emerging technologies to promote transparency along food supply chain.

Communication

Like other countries in arisen area, China has put emphasis on the importance of labeling, risk communication to promote product, and production transparency characteristics.

Labeling

According to Chinese Food Safety Law, requirements for labels, marks, and instructions related to health, nutrition, and other food safety requirements should be prescribed by national food safety standard, which is the standard for mandatory execution. This is also the General Standard for the Labeling of Prepackaged Foods (GB 7718). Nowadays, the national food safety standard is under revision. Newly updated rules include the changes in indicating allergens from voluntary presentation to mandatory presentation. Meanwhile, it will also introduce deregulation on the information about food quality from mandatory requirement of grading information to a recommended one. Notably, only information about food safety can be required in the form of the national food safety standard. Therefore, food business operators or local governments can take advantage of industry standards or group standards to present quality information to meet consumers' demands of diverse choices. A typical case is the Shenzhen Standard provided by city of Shenzhen as special economic zone. In the case of food quality, such local standards are aimed to provide local consumers with safer food of high quality.

When it comes to labeling information, the arrangement of punitive damages has increased consumers' awareness to take advantages of such information to make an informed choice. As is permitted under the law, consumers are entitled to require a producer of food failing to meet the food safety standards or a trader knowingly dealing in such food to pay an indemnity of ten times the price paid or three times the loss; or if the amount of the additional compensation is less than 1,000 yuan, it shall be 1,000 yuan. Comparatively, it is much easier to judge whether a food is failing to meet the food safety standards or not by reading labeling information, such as missing producer information or date of production. While so-called professional consumers target labeling mistakes having nothing to do with food safety issues for economic gain, an exception in the law in which a defect in the labels or instructions of the food which neither impairs food safety nor misleads consumers is exempted from punitive damages.

Grading system

Like Singapore, the above-mentioned grading system has also been introduced to food safety regulation for catering services in China. As required by the Guiding Opinions on the Implementation of Quantitative and Classified Management of Catering Services provided by the former Chinese Food and Drug Administration (FDA), such grading system is applied to restaurants, fast food shops, snack bars, drink shops, canteens, collective dining distribution unites, and central kitchens. Using requirements from the Food Safety Law, the Implementation Regulations of the Food Safety Law, the Measures for the Administration of Catering Service Licenses, and alike as reference, the evaluation is based on the license management, personal management, site environment, facilities and equipment, procurement and storage, processing and manufacturing, cleaning and disinfection, food additives, and transportation inspection. The result of such evaluation is graded in two ways. One is the dynamic grading to present the status of the catering service confirmed by official control and divide into three levels of excellent, good and general, which are represented by three cartoon images: laughter, smile, and straight face. The other is the annual grade, which presents the comprehensive evaluation of the results of the official control during the past 12 months and is divided into three grades of excellent, good, and general in the form of A, B, and C, respectively. Accordingly, consumers may pick up qualified restaurants with grade of Laughter and A to encourage the food operators to take care their internal management according to food safety requirements. Notably, with such guidance at national level, it relies on the local competent authorities to specify the grading system when it comes to regulating catering services. As a result, the images for grading are still different from region to region.

In addition to legally required information disclosure, official control in China has paid more attention to name and shame list for public review in the form of so-called credit system, which can further improve food operators' compliance awareness due to the consumers voting with their feet and popular opinion about their reputation, thanks to the accessibility of operators' compliance or noncompliance records. As a matter of fact, the building of social credit system is accelerating in China in the wake of the construction of Credit China. Indeed, more people have begun to take their credit seriously given the convenient services resulting from good credit, or punishment from bad credit. In view of this, the credit system which was originally widely used in the financial sector has been extended to other fields, including food safety regulation, with the purpose of building a credit-based sanction in both a positive and negative way. Accordingly, the former Chinese FDA has issued guidance and department rules for specifying the roadmap to build a food safety credit system. Based on this, local authorities have also explored such a specific credit system to improve food safety by combining the national requirements and local specialties. However, during the latest round of administrative reform, the independent model of food safety regulation in form of the Chinese FDA has been replaced by a comprehensive model of State Administration of Market Regulation. As provided by Implementing Regulation on Food Safety Law in 2019, the new department shall establish a credit supervision in conjunction with relevant departments, in order to provide joint rewarding and punishment mechanism. Based on the combined credits files, a blacklist shall be set up to punish serious violators and restrain their activities in the field of market access, financing, etc.

Risk communication

Provided as legal principle, governance has been a paradigm to ensure food safety by addressing that food safety is everyone's business. To this end, risk communication has been introduced into the revised Food Safety Law in 2015. Based on this, competent authorities, scientific experts, and the media are actively engaging in risk communication with the public for disseminating real food safety knowledge and facts. For example, either the former Chinese FDA or current State Administration for Market Regulation (SAMR) has put a column on their official website providing information about risk alert and communication, such as the risk for a chemical or biological hazard if it is found during the national sampling and testing examination.

Also, a challenge for food safety governance in China is the repeated food safety rumors over these years, which not only deteriorate the economic environment for business development, but also increase the difficulty in recovering the public confidence in both business and official control. As stated in a National Notice on Strengthening the Prevention, Control and Governance of Food Safety Rumors, information disclosure is a key weapon to curb the spread of food safety rumors while stakeholders should work together. Among the others, it is the company involved in the rumor that assumes the primary responsibility to inform the public of the truth.

Technology-based information regulation

As mentioned in a statement on steps to usher the United States into a "new era of smarter food safety" provided by commissioners from FDA, one of characteristics in the new era of smarter food safety policy framework is to use emerging technologies to create a more digital, transparent, and safer food system.[8] Nowadays, both self-regulation and government regulation have also become "smarter" by integrating these new technologies into risk management, including the practices in China.

Official control in a smarter way

In China, official control has been carried out in a smart way by taking advantage of mobile internet, big data, and artificial intelligence (AI), as in the case of online administrative licensing, daily supervision, sampling, and testing at random. This shows the willingness of the Chinese government to update the regulation in terms of technology advancement, business development, and capacity building. Thus, regulatory improvement has been present in the form of more responsive and effective regulation. Correspondingly, with the support of technology, governments at all levels and competent authorities in China are actively exploring advanced models of smart regulation that combine various emerging technologies to adapt to new developments and change traditional regulatory models.

Taking sampling and testing system as an example, it has been taken as an important regulatory approach by the former China FDA since the revision of Food Safety Law in 2015. Generally speaking, organized in a regular basis, unannounced national sampling and testing are carried at local levels while the results are published for public review. Additionally, foods failing to meet food safety standard found by this national inspection are subjected to food recalls and official investigation if there are violations. In this aspect, the "Administrative Measures for Food Safety Sampling Inspection" stipulates that the regulatory authorities shall publish information on unqualified food safety inspections, including the names, specifications, production dates or batch numbers, unqualified items, the names of producers, the trademarks, address, etc. Therefore, such official control is aimed to monitoring food business operators' fulfillment of their primary responsibility, to guide consumer's information-based consumption and to promote governance for food safety.

As far as smart regulation is concerned, big data has already existed after several years of collection, which can be further analyzed to reflect national and regional food safety statues and predict sectors with high risks. For example, the annual result of national wide sampling and testing system has increasingly become an important indicator for confirming China's high protection level in the case of food safety regulation. Besides, to protect the right to know of different stakeholders, the former China

---

[8]Statement from Acting FDA Commissioner Ned Sharpless, M.D., and Deputy Commissioner Frank Yiannas on steps to usher the United States into a new era of smarter food safety, April 30, 2019, available at: https://www.fda.gov/news-events/press-announcements/statement-acting-fda-commissioner-ned-sharpless-md-and-deputy-commissioner-frank-yiannas-steps-usher

FDA developed a mobile application called "Food Safety Check" to provide available results from sampling and testing based on the publication columns and query databases specified for sampling inspection information on official websites. At present, consumers can obtain sampling information through multiple channels (such as mobile phones, websites, newspapers, and televisions), which enhances the accessibility of information and enhances the public's sense of shared regulatory results. In addition, based on the random inspection work, there are also food safety risk communication and early warning. In this basis, taking into account the current emergence of new food business models, food safety asymmetry and inadequate communication of food safety risk cause the internal office for sampling and risk communication inside SAMR to shift preparation supervision activities from reactive to preventive. To this end, it is necessary to gradually build an early warning indicator system and analysis system, information release system, response system, and re-evaluation system for food safety risks with Chinese characteristics, in order to establish a unified food safety information release platform, and improve and build an early warning communication system suitable for national conditions. As far as progress is concerned, the "Design Scheme of Food Safety Risk Early Warning and Communication System" prepared by the Chinese Academy of Inspection and Quarantine will provide theoretical and path guidance for the next work.

High-tech self-regulation

For retailers, food safety at their own jurisdiction still depends on the fulfillment of responsibility of suppliers in addition to their own efforts. In view of statutory obligations and consumers' expectation in this regard, the supply chain management carried out by retailers includes the safety management along the supply chain. Among the others, food safety traceability serves as a tool to collect and share information about products and suppliers, which provides basis for consumers' informed choices. More importantly, once a food safety problem is discovered, the information can be used to ensure that the product at issue can be recalled, the cause can be investigated, and the responsible operator can be located. Comparatively speaking, establishing a food safety traceability system by means of technologies is the embodiment of the operators' willingness of fulfillment of obligations to ensure food safety. Nowadays, in China, the trend of "everything on the blockchain" will undoubtedly bring first-mover advantages such as market opportunities and reputation to pioneers who take advantage of newly emerged technologies such as blockchain to reinforce food safety traceability.

The application of blockchain technology runs through the entire food supply chain, and information can include the collection of information of agricultural inputs at the primary stage and consumer information at the retail stage, as well as transportation, storage, and other information in the process. By ensuring the authenticity of the information on the chain and the convenience of information acquisition and sharing, participants in the chain can not only optimize self-management and forecast and plan business progress in advance, but also strengthen coordination and cooperation with partners. In particular, in the development from material scarcity to overproduction, customer demand-oriented supply chains can promote supply-side reforms through precise matching, integrating resources, and optimizing processes, thereby ensuring the quantity and quality of food to satisfy consumer demand and establish a cooperative development mechanism for win-win cooperation between upstream and downstream operators. Moreover, from traceability information to a traceability platform based on the blockchain, the participation of stakeholders and the governance of the alliance of private operators have also fulfilled the principles of social co-governance in food safety.

It is true that the introduction of new technologies will increase short-term management costs, but such burden-increasing options can take advantage of supply chain visualization and smart management to improve the supply chain management efficiency, cater to industry development trends, and respond to consumer demands. For example, increasing transparent management of the supply chain is a growing regulatory requirement and consumer expectations. Therefore, for one thing, business operators are also willing to "do the right thing" to promote their sustainable development, as in the case of apply new technologies. For another thing, government regulation also needs to keep pace with the times, and then provide a suitable "deregulation" environment for new development and innovation. Here, the "Guiding Opinions of the General Office of the State Council on Actively Promoting the Innovation and Application of Supply Chain" pointed out that the orientation of the supply chain is an organizational form of efficient coordination throughout the process, and the integration of resources by the guidance of customer needs. Therefore, when new information technologies are integrated in supply chain management, innovations in industrial organization, business models, and government governance are also the components of supply chain innovation and application scenarios. In other words, technologies such as blockchain are not just new productivity in the Internet era, but also new production relationships, which are synergistic and mutually beneficial.

In practice, the introduction of blockchain technology can be initiated by a single party such as a retailer, relying on the technology supply of a third party. However, for the participants joining the chain, the relationship between them is equal, including the sharing of responsibilities in terms of information recording and storage. Similarly, the choice of management model is based on the participants' prior consent or mutual discussion. During this period, online and offline data synchronization to ensure the authenticity of the data at source is an important management principle, and it is required that the information be uploaded to the chain as soon as possible after it is generated. When online performance can rely on technical regulation, whether the offline synchronization is performed, that is, whether to fulfill the data upload agreement or how to punish the violators, the market can play an important role on its own. For example, a third party can be jointly selected to audit the performance of each participant, delisting from the supply chain can be used as a punishment, and product promotion can be used as an incentive for participation and cooperation. Of course, greater success comes from consumer recognition.

However, even if food safety traceability has become a popular application scenario of the blockchain, the operation for food production, distribution, and sales are still offline for businesses. As a matter of fact, there were also information technologies such as QR codes for food identity verification and traceability before the introduction of blockchain technology. In the face of offline logistics and online information flow, it cannot dispel the question of whether blockchain technology can truly synchronize offline/online information as long as manual information entry is still part of this process. The constant emergences of "blockchain-based rice" and "blockchain-based liquor" have also exacerbated the chaos in the application of blockchain technology to food safety. Therefore, the innovative and cooperative governance promoted by the application of blockchain technology introduced by food business operators is still needed to be guided and even regulated by the government in a cooperative way. Taking China as an example, government regulation is shifting from "absent" to "returning" with regard to the expanded application of the new technology of blockchain. From department rules to international technical standards, the government's involvement is not only to strengthen the regulation of blockchain technology, but also to use blockchain technology to achieve technology-driven regulation. For the former, the "Rules on the Management of Blockchain Information Services" issued by the Cyberspace Administration released regulatory signals intended by the relevant national authorities to oversight the application of blockchain technology through the refinement of rules. For the latter, the concept development of "tech-regulation" and "tech-governance" is to promote a new technology-centric regulatory paradigm, which includes a data-driven regulatory upgrade to carry out real-time monitoring of events to detect and dispose of illegal acts in a timely manner, and to promote the interaction of data between public and public entities and public and private players. To this end, the standardization of blockchain technology is the basic work to promote consensus and data sharing among different stakeholders.

The role of platforms

In the so-called Internet ecosystem, cross-border, collaboration, and integration have become the characteristics and highlights of Internet technology to connect everything. For example, while online platform for meal ordering is supposed to develop as a comprehensive platform for suppling all kinds of service for local life, it is also vertically focused on deepening online catering services, and horizontally continues to expand new businesses such as bicycle sharing, hotel accommodations, and so on. In order to ensure the inclusiveness of this digital development, such as the rights of online business operators and consumers, the development of e-commerce for both inside and outside services, the legal environment for e-commerce is also being continuously optimized. This includes the promulgation of general laws such as the "Food Safety Law" and other general laws such as the "Internet Security Law" and "E-Commerce Law," as well as the adjustment of regulatory system such as comprehensive regulatory model for market regulation and co-governance. For instance, Article 7 of the "E-Commerce Law" stipulates that the state establishes a collaborative regulatory system in line with the characteristics of e-commerce and promotes the formation of a governance system for e-commerce markets that is jointly participated by relevant departments, e-commerce industry organizations, e-commerce operators, and consumers.

In a governance system for e-commerce market, the government has the dual role of regulator and facilitator. In order to play this dual role, co-regulation between government and business operators has become an important option for co-governance to guarantee food safety. For example, in 2015, the original Shanghai Food and Drug Administration took the lead in issuing the Guidance on Encouraging Online Third-Party Platforms for Meal Ordering to Collect and Apply Government Food Safety Data, encouraging the them to collect and apply the data collected by government during the official control of licensing and supervision, and encouraging the incorporation of supervision information into their own credit evaluation system, which laid a foundation for the platform to better fulfill its obligations such as license review, and improve the accuracy and effectiveness of such review. Also, in 2018, the Jilin Provincial Market Regulation Department and a platform provider signed a strategic agreement to promote cooperation. According to this agreement, the two parties will carry out in-depth cooperation in strengthening information sharing of online food business operators, establishing a cooperation mechanism for consumer rights protection, commodity quality control, and jointly regulating illegal acts. For local governments, this kind of cooperation is not only for upgrading tech-regulation, but building a "Jilin brand" of co-regulation between government and business operators. It also embodies the emphasis on the digital economy and smart regulation, as well as determination to create a supportive and evolving regulatory environment for business and economic development. In order to achieve this co-regulation, the technical and data capabilities of the online platform will play a key supporting role.

Taking the Jilin case as an example, the data provided by platform includes the data of online food business operators and the quality of their goods, negative evaluation information of these operators and complaints raised by the consumers. With these data, regulators can strengthen regulatory targeting and preventive measures based on aggregated issues and risks. Comparatively speaking, the innovation of this cooperation lies more in the upgrade from static information reporting to dynamic information sharing, so as to make platform data available for official control. In this aspect, the "E-Commerce Law" itself provides requirements for government data acquisition and data notification at the side of platform. For example, if the relevant competent authority requires e-commerce operators to provide information about e-commerce data in accordance with laws and administrative regulations, such operator shall provide it. However, the current legislation regarding internet regulation is still in the process of improvement, as the laws and administrative regulations on government data acquisition mentioned in the above clauses need to be further clarified. With these more specified requirements, platforms may shift from passive cooperation to active cooperation and thus promote a co-regulation or governance attitude of "solving problems rather than creating problems." For example, through shared information about food safety risks in official control, cooperation can be enhanced between government and citizens to prevent intentional or unintentional public health risks.

Notably, the so-called data sharing or exchange, which focuses more on the two-way information communication, that is, platform or other stakeholders also need data open at the side of government to strengthen their management. Moreover, when data becomes the new factor of production, the openness of the government's own data can also provide data support for corporate innovation and help the development of the digital economy. However, it needs to be pointed out that the government has both the obligation to protect the security of information and the obligation to keep personal information, privacy, and business secrets confidential. Comparatively speaking, the disclosure and sharing of basic regulatory information, such as business licenses, food business licenses, and registration records, is necessary and important to improve the level of food safety governance, and it is the basic guarantee for online and offline collaborative governance of food safety.

Conclusion

As mentioned in the previous section, information flow along the food supply chain can be provided by competent authorities or food business operators according to the different regulatory tools. Legally speaking, it corresponds to the government's duty or operator's obligation to disclosure information, as in the case of administrative information disclosure or labeling. Nowadays, newly emerged technologies not only change the way how food businesses and official controls are carried out, but also the way how information is provided. When such development considerably contributes to transparency along food supply chain, food safety regulation between regulator and regulatee may shift to food safety governance in which more parties have the opportunity to play a role in ensuring safety.[9] At this point, the Chinese experience can be regarded as a model to put the so-called "food safety is everybody's business" to practice, which was emphasized by the World Health Organization and United Nations during the first World Food Safety Day in 2019. In fact, it has been a common recognition in which experiences have been "imported" or "exported" among different Asian countries, such as the raking system for restaurants in Singapore and China, or how South Korea introduced punitive damages like China during the revision of product liability act.[10] Undoubtedly, these experiences can continually improve food safety at international levels with ongoing exchanges around the world.

[9] Lepeintre J., Sun J. (eds), 2018. Building food safety governance in China, Publications Office of the European Union, Luxembourg, available at https://eeas.europa.eu/sites/eeas/files/building_food_safety_governance_in_china_0.pdf

[10] Korean Product Liability Act Amended to Include Punitive Damages and A relaxted Burden of Proof, available at https://www.thekoreanlawblog.com/2017/06/product-liability-law-korea.html.

chapter ten

Food labeling in Africa

Wele Elangwe, Rosemary Agbor, and Fabrice Mbala

Contents

EDITORS' NOTE: FULL-CIRCLE REGIONAL FOOD SYSTEM TRANSPARENCY WITH A FOCUS ON AFRICA

Last but most certainly not least, three renowned experts comment on information flow in the food system focused generally on the West and Central regions on the African continent. The perspective sin this chapter provide a most insightful comparative study that informs many aspects of information paths in the food system. CEMAC (Commission of the Economic and Monetary Community of Central Africa) region and the Organization for Harmonization of Business Law in Africa (OHADA), and the Africa Continental Free Trade Area (AfCFTA) are rooted in civil law and provide a perspective that is quite complimentary to those in Part II of this book. While the following chapter joins with the chapters on the European Union (Chapter 8) and China (Chapter 9) in discussing civil law regulation, the elements of information sharing across food supply chains has similar roots and challenges.

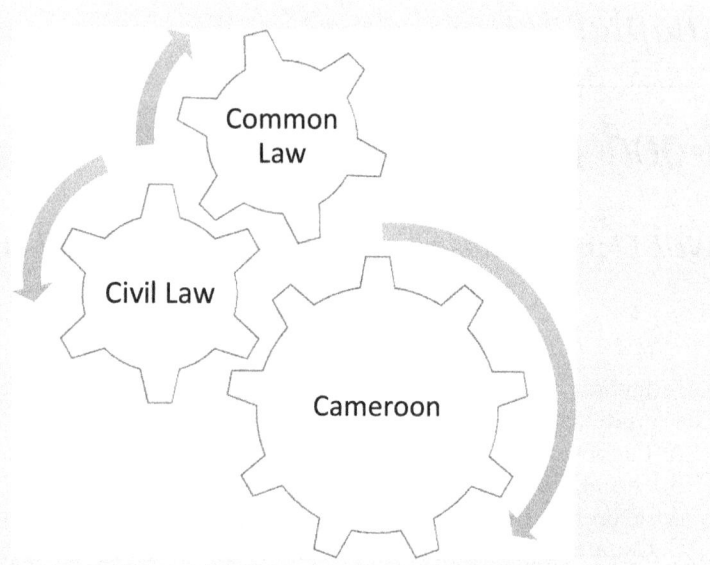

Figure 10.1 Elements of both common and civil law flow into the food regulation in Cameroon.

This chapter's case study on Cameroon examines commonalities in information sharing that relates to the themes of this volume: food safety, public health preservation, consumer rights, labeling, misbranding, and supply chain management. Harmonization of rules and the free movement of goods is just as important in Africa as it is in other continents. But, regional legal approaches within Africa differ from European, Asian, and North American approaches. For this purpose, the case study on Cameroon is placed last because it relates both civil and common law elements to the shared subjects of all preceding chapters.

Several recurring topics from this book come together in the final chapter to drive home themes of food system transparency around the world and provide invaluable lessons when viewed from various perspectives.

Introduction

Food transparency is an essential step in the global transformative food policy system intended to reduce health disparities and raise health standards for all irrespective of country or continent. While Africa needs to take several steps to boost food transparency, it is important to recognize that there are no common food standards frameworks or authorities on African food labels. Thus, food transparency and food safety requirements vary from country to country. Specifically, with most African countries now embracing international trade, much more attention is being given to food transparency than before. Because of this interest in international trade, trading blocs are sprouting and easing trade within and outside Africa which, in turn, is increasing the risk of the spread of negative effects from inadequate food transparency regulations. Transparency in the food policy system is a key focal point of efforts to improve health by providing consumers with necessary information to make good nutritional choices, as well as to achieve sustainable food chains and ensure food safety and quality. Hence, it is critical to clearly regulate food labeling in order to ensure resultant food transparency. Food labels are important not only to promote and advertise products but also to provide important information for consumers on what they are ingesting. They contain nutritional information, information on allergens, and help individuals take stock of their daily dietary requirements.

Compared to other African countries and more generally to its neighboring countries of the CEMAC (Commission of the Economic and Monetary Community of Central Africa) region, Cameroon's industrial environment is relatively diversified with a large number of informal businesses.[1] Cameroon is also bilingual with both French civil law and English common law in effect as legal systems making it possible to glean similarities with countries sharing a similar colonial past be it from colonialism by Britain or by France. Specifically, business law is being harmonized in Cameroon through the Organization for Harmonization of Business Law in Africa (OHADA), which draws heavily on French civil law. OHADA was created by a 1993 treaty, comprising of seventeen member states,[2] and its uniform acts are directly applicable in each member state, resulting in legal harmonization in Francophone Africa.

Geographically, Cameroon forms a bridge between West Africa and Central Africa. Cameroon is bounded on the east by Chad and the Central African Republic; on the south by the Republic of the Congo, Gabon, and

---

[1] https://www.iso.org/files/live/sites/isoorg/files/archive/pdf/en/2012_economic_benefits_of_standards_2_cameroon_chococam_en.pdf

[2] OHADA's member states are Benin, Burkina Faso, Cameroon, Central African Republic, Chad, the Comoros, the Republic of Congo, Côte d'Ivoire, Equatorial Guinea, Gabon, Guinea, Guinea-Bissau, Mali, Niger, Senegal, Togo, and the Democratic Republic of Congo. These countries have adopted French as one of their official languages, except for Guinea Bissau, which is Portuguese speaking.

Equatorial Guinea; and on the west by the Atlantic Ocean and Nigeria.[3] Reviewing food transparency regulations in Cameroon therefore provides a broad perspective on food transparency implications for West and Central Africa.

With the recent launching of the operational phase of the Africa Continental Free Trade Area (AfCFTA), which is expected to go into effect in July 2020, understanding food transparency and promulgating regulation becomes even more critical in Africa. The goal of the AfCFTA among others is to establish a single market for goods and services across all fifty-five African countries and allow the free movement of business travelers and investments in the continent. Except for Eritrea,[4] all remaining fifty-four African countries have either signed or ratified the Free Trade Agreement.[5] Therefore, trade will become even more seamless and lucrative, making the need for food transparency regulation even more dire.

Case study: Cameroon

In Cameroon consumer knowledge, use, and understanding of food labeling has not been investigated extensively. With prevailing high levels of poverty, people eat and drink what they can afford and have available with little or no regard for nutritional content. Very few people bother to know if what they buy from grocery stores, local shops, and off licenses are safe, healthy, or fit for human consumption. Even many more are unaware of the rights they have as consumers to call to order or report the lack of duty of care by the huge multinationals supplying beverages and other prepackaged foods, when the latter fail to uphold basic regulations and legal standards of compliance for the supply of products for public consumption. This perhaps could be one of the contributors to the country's unusually high mortality rate estimated to be at 342/1000; almost two times the global average of 178/1000.[6] Nevertheless, the ravaging effects of such alarming rates can only be considerably curtailed or controlled through the promotion of healthy eating habits, and more especially through the provision of adequate nutritional information on food labels.

According to the World Bank's most recent economic plan, Cameroon shall become an emerging economy by 2035.[7] However, it is only

[3] http://www.intracen.org/exporters/organic-products/country-focus/Country-Profile-Cameroon/
[4] https://www.tralac.org/resources/by-region/cfta.html#legal-texts
[5] Id
[6] https://data.worldbank.org/indicator/SP.DYN.AMRT.MA?locations=CM
[7] https://www.prc.cm/en/the-poles/economic-emergence-action

reasonable to expect that this economic plan would have to pass through a certain degree of economic transparency requiring adequate measures to reinforce the country's economic competitiveness, by supporting the development of standards systems to assist businesses in their quest for certification.

In that light, there have been many developments in Cameroon in terms of food labeling in the last decade. First is the enactment of Law No. 2018/020 of 11 December 2018, which is the framework law on food safety in Cameroon. Second, the creation of the Standards and Quality Agency of Cameroon better known by its French acronym (ANOR). ANOR is a governmental organization tasked with instituting a resourceful quality control infrastructure in which standardization has a fundamental function. As a member of the International Organization for Standardization (ISO),[8] ANOR has carried out various studies using the ISO methodology to determine the benefits that standards have brought in terms of higher quality to the manufacturing of prepackaged food products. For instance, in 2019, ANOR conducted a case study on Chococam, a chocolate manufacturing company in Cameroon, to determine the benefits that standards have brought in terms of reduced costs and higher quality products.[9]

Although the food industry in Cameroon and consumers are facing many challenges with regard to food labeling, some organizations have taken it up to promote the use and understanding of nutritional information on labels to consumers. Notably, nongovernmental organizations, such as *La Fondation Camerounaise des Consommateurs* (FOCACO; loosely translated as the Foundation of Cameroon Consumers), are very active in promoting the rights of consumers and hold food manufactures accountable for transparency with respect to food labeling and specifically to nutritional contents of their products.[10] Although the enactment of this regulation may not be a panacea to the deficit of information on food safety and food labeling in Cameroon, it has provided a level ground between consumers and food manufacturers with regard to their respective rights and obligations. Manufacturers are now aware of the duty of care owed to consumers in compliance with domestic laws, as well as international norms and quality standards.

[8] At the end of 2018, Cameroon was among the 162 countries who were part of ISO's worldwide membership, with ANOR specifically falling within the ranks of the 120 national standards bodies (NSBs) that were full members. 39 NSBs were correspondent members and 3 NSBs were subscriber members. See ISO 2018 Annual report, available at https://www.iso.org/publication/PUB100385.html

[9] https://www.iso.org/publication/PUB100385.html

[10] https://www.facebook.com/ayissiabena/

A. Purpose of food labels

In Cameroon, the Biosafety Law No 2003/006 of 21 April 2003,[11] was one of the country's first attempts at regulating food products, additives, and supplements intended for human consumption, in order to ensure the protection of consumers, promote health and safety, as well as the protection of the environment. The law, structured to cover various stages of the food chain from production to manufacturing, also covered preparation, handling, parceling, storage, transportation, packaging, preservation, importation, exportation, distribution, and sale of food products. This legislation came to breach the gap created by the growing need in Cameroon for more up-to-date food labeling regulation to better protect and inform the consumer; to prevent and control the impact of noncommunicable diseases (NCDs); and to align with new, emerging scientific nutrition-related research, new trends, and international standards and guidelines, including the *Codex Alimentarius*.[12]

For purposes of this chapter, food will be defined as "any substance, whether processed, semi-processed or raw, which is intended for human consumption includes drink, chewing gum and any substance which has been used in the manufacture, preparation or treatment of 'food' but does not include cosmetics, tobacco or substances used only as drugs."[13] In 2018, parliament promulgated Law No. 2018/020 of 11 December 2018 to further regulate food safety within its territorial boundaries. This law defined a label as "any indication, mark, sign or design or any other written description, printed, painted, marked, etched or stamped on the food packaging or any other feature related to packaging." The main purpose of food labels is to inform consumers and to provide strategic support in the commercialization of the product. According to the *Codex general standard for the labeling of prepackaged foods*, food labeling includes "any written, printed or graphic matter that is present on the label, accompanies the food, or is displayed near the food, including that for the purpose of promoting its sale or disposal."[14]

---

[11] https://www.acbio.org.za/en/explanation-and-comments-cameroon-biosafety-law-mariam-mayet-april-2004

[12] The Codex Alimentarius, or "Food Code" is a collection of standards, guidelines, and codes of practice adopted by the Codex Alimentarius Commission. The Commission, also known as CAC, is the central part of the Joint FAO/WHO Food Standards Programme and was established by FAO and WHO to protect consumer health and promote fair practices in food trade in 1963. See http://www.fao.org/fao-who-codexalimentarius/en/

[13] See Definitions for the purposes of the Codex Alimentarius, http://www.fao.org/3/w5975e/w5975e07.htm

[14] The Codex General Standard for the Labelling of Prepackaged Foods was adopted by the Codex Alimentarius Commission at its 14th Session, 1981 and subsequently revised in 1985 and 1991 by the 16th and 19th Sessions and amended by the 23rd and 24th Sessions, 1999 and 2001. This standard has been submitted to all Member Nations and Associate Members of FAO and WHO for acceptance in accordance with the General Principles of the Codex Alimentarius. Available at http://www.fao.org/3/Y2770E/y2770e02.htm

Food labels greatly aid in preventing the consumers from getting confused with respect to the quality and content of the products they intend to consume. In this regard, it is important to provide the consumer with all the necessary information on the ingredients, nutrition, and the declaration of potential antigens and foreign substances, and/or health hazards in order to help consumers make a conscious choice based on an informed decision. More recently, the government of Cameroon has placed so much importance on food safety and consumer protection at the heart of her food policy. This is especially so because food labeling is considered to be a valuable and relatively low-cost tool in the prevention of NCDs and promoting health and wellness for all.

Since food labeling is a global trend, the Cameroonian legislator recognized the need to ensure that all food-related activities are executed in accordance with the provisions, standards, guidelines, and other national and international recommendations contained in regularly ratified international treaties governing food safety. Notable regulatory instruments considered include Codex Alimentarius, World Trade Organization (WTO), International Plant Protection Convention (IPPC), World Organization for Animal Health (OIE), Worldwide Organization of Consumers (OIC) as well as those established by the Cartagena Protocol on Biotechnological Risk Protection. Likewise, the promulgation of Law No. 2018/020 of 11 December 2018, especially Article 8, has also been another attempt at regulating Food Safety in Cameroon. Per this article, the food legislation shall seek to protect the interest of consumers and provide them with essential conditions for making a well-informed choice of food products to consume. In this regard, it is responsible for preventing fraudulent and misleading practices, the marketing of defective or adulterated food products, and any practice that may mislead the consumer.

B. The significance of food labeling during consumer decision-making

In Cameroon, food purchases are commonly seen as routine purchase decisions which require little involvement and no external search for information. Thus, very few Cameroonian consumers pay any attention to label information. The few who do so study the labels of food products with more complex nutritional composition more carefully than products with which they are more familiar, or which they find easy to interpret.

Consumers behave and make decisions in different ways for different reasons. A complex combination of external and internal factors influence consumers' food product-related needs. These include various demographic characteristics of the consumer, such as age, gender, education level, race, ethnicity, income, work status, and product knowledge, needs, personality, hunger, and marketing-related influences. This explains why Section 11 of the law regulating food safety in Cameroon enjoins business entrepreneurs involved in the food sector to make sure that they are compliant with governing food legislation in the course of their activities and insists on the fact that they shall have the direct responsibility of protecting the health of consumers in relation to the products they release for consumption.

Food labels are particularly important when addressing consumers' needs, while food packaging, which often integrates labeling information as part of the container, can influence consumer purchasing behavior. Elements of food packaging include package color, image, typeface, and type of packaging, and can generate an emotional response in consumers.

C. Cameroon consumer food preferences and their labeling

Cameroonian consumers have a huge preference for imported food products despite the products' high price tag. Low domestic production/value-added processing has made the local populace shun locally manufactured products as lower quality items. This is abated by the government who spends millions of CFA franc in food import subsidies to avoid a recurrence of the 2008 food riots. Food import subsidies account for approximately 20 percent of the current national budget which is highly dependent on increased commodity exports.[15]

Most of the products available in the food section are originally from France or the European Union (EU), with some American-branded products such as Kellogg's and Kraft Heinz also serving as local favorites. Common consumer-oriented products include, among others, tomato ketchup, mayonnaise, salad dressing, canned soups, vegetables, fruits, baby foods, health food products, powdered milk, rice, fish products, wheat, foreign wines, liquor, and spirits. Therefore, the question as to the labeling of these wide variety of food products imported into Cameroon yearly must be highlighted, as they have a direct impact on local consumers in relation to what they purchase and consume.

To that effect, ANOR sets and enforces mandatory standards on food packaging and labeling. It also enforces international standards established by the Codex Alimentarius Commission and other international

---

[15] Marcela Rondon, *2013 Exporting to Cameroon*, GAIN (Global Agricultural Information Network) Report of 14/3/2303 at page 5.

norms. According to a 2013 Global Agricultural Information Network (GAIN) report, there are specific labeling mandatory standards that must be incorporated for imported food products.[16] It specifies that labels must be in English and French for easy identification.[17] Also, for the purposes of custom clearance, proper labeling is pivotal as the agencies in charge of the control of imported product working hand in hand with ANOR may have to inspect the quality of the food products shipped into the country for conformity with the mandatory standards.

Food products imported into Cameroon must state their manufacturing and expiration dates engraved on top of the container and packaging in legible ink. It must also indicate the country of origin and a "made in" and "to be consumed before…" date.[18] These fundamental standards cannot be overlooked as they go a long way to guarantee the safety of consumers and to ensure that food products imported to Cameroon for consumption are fit for that purpose in order to avoid adverse health effects.

D. Information on food labels imported or manufactured in Cameroon

The intention of the new legislation, Law No. 2018/020[19] of 11 December 2018, regulating Food Safety in Cameroon, is to close known gaps which may permit confusion in labeling and control, and to ensure that consumers have access to quality food products. The promotion of a healthier eating environment through improved labeling and control was certainly one of the key objectives behind these regulations to promote better food choices and improve public health and safety.

According to this law, every food product in Cameroon must now contain:

- Name and brand of product
- Country of origin
- Name and address of the manufacturer
- List of ingredients (including salt) and additives
- Nutrition and caloric value
- Allergenic ingredients and allergenic processing aids (gluten, soybean, peanuts, etc.)
- Net weight/volume (metric system)

---

[16] Id at page 9
[17] Id
[18] Id
[19] https://www.prc.cm/en/news/the-acts/laws/3232-law-no-2018-020-of-11-december-2018-framework-law-on-food-safety

- Expiration date/shelf life
- Shelf life (if the product has a shelf life of less than three months, it must include the day/month/year of expiration)
- Products containing GMOs must be labeled accordingly
- All products sold in Cameroon must have the following statement *"Sold in CEMAC"* or *"Vente en CEMAC"*

Since the passage of this law, a higher burden of care has been placed on food manufacturers and distributors. For example, following Section 24 of this law, all economic operators, as well as food-producing companies, must comply with the strict rules of hygiene in line with the standards of Codex Alimentarius or the guidelines of any other international organization involved in food safety. Section 26 likewise mandates that all prepackaged foodstuff be labeled in accordance with the legislation in force. Therefore, as a general principle which should apply in Cameroon, prepackaged foods should not be described or presented on any label or in any labeling in a manner that is false, misleading, or deceptive, or likely to create an erroneous impression regarding its character to the consumer.

For the most part, the majority of food products imported into Cameroon and sold in open markets and other supermarkets easily conform with these fundamental labeling standards. But the challenge remains with those locally involved with the production, manufacturing, preparation, manipulation, treatment, packaging, transportation, packing, conservation, importation, exportation, distribution, and sale of food products. A case study is in the labeling of alcoholic and nonalcoholic beverages. Some of the products sent to the market for mass consumption are not compliant with the required standards and norms for labeling of prepackaged food products. A glance on their packages do not reveal many relevant information necessary for the consumer to make conscious decisions as to what they choose to consume and the effects and consequences it may have on their health.

In January 2017, for example, the Fondation Camerounaises des Consommateurs (Cameroon Consumers Foundation) reported that the Cameroon National Brewery Company (Les Brasseries Du Cameroun) failed to uphold the standard for the labeling of alcoholic beverages in Cameroon.[20] They indicated that new locally brewed beer "Manyan" did not meet up with the standards by failing to mention the percentage of alcohol on its label, in accordance with NC 208: 2003-02 (drinks containing wheat) and CN 04. 2000-20 (labeling of prepackaged foods).[21]

---

[20] https://actucameroun.com/2017/01/22/cameroun-la-nouvelle-bire-manyan-est-un-pige-absence-du-degr-dalcool-sur-ltiquette/
[21] Id

Source: http://www.camer.be/57423/1%3A11/cameroun-la-nouvelle-biere-manyan-ne-repecte-pas-la-norme-absence-du-degre-d39alcool-sur-l39etiquette-cameroon.html

They went further to indicate that by launching the alcoholic drink called "Manyan: Beer of the Country" without the degree of alcohol on the label, the National Brewery Company manifested a total lack of duty of care and failed to uphold the standards of required transparency toward consumers.[22] In the absence of such relevant cautionary information, the health and safety of the consumers were put at risk. Recourse was therefore taken to the government to demand the recall of all noncompliant Manyan products; and to enforce the National Brewery's adherence to all labeling requirements as prescribed by law.

[22] Id

Le Brasseries du Cameroun – Manyan Beer

Source: https://untappd.com/b/les-brasseries-du-cameroun-bgi-douala-manyan/
2059305/photos

Another Cameroonian company that has come under the scrutiny of the
Cameroon Consumer Foundation (Focaco) for failing to maintain the food
labeling standards enforced in Cameroon for their food products in the
market for public consumption is *Chococam*, a local chocolate company,
and subsidiary of South African food giant, Tiger Brands.[23] In September
2019, *Chococam* was embroiled in a scandal following doubts if there were
any chocolate nutrients found in their prepackaged chocolate pastes,
"Tartina" and "Choconut."[24] *Focaco* alleged that they had conducted a survey which answered the above questions in the negative. They issued a
statement to the effect that these spreads indicate chocolate on their labeling whereas they do not contain any at all.[25] The report of the foundation
disclosed that the company was using the terms "choco" and "chocolate
taste" to advertise the products on the labels of their packaging when they
contain at most cocoa powder.[26]

Speaking as president for *Focaco*, Alphonse Ayissi Abena argued that
the label "Tartina" affixed on the packaging mentions "cocoa powder,"
while the spread contains more peanuts than cocoa powder. This was a

---

[23] https://www.investiraucameroun.com/gestion-publique/1709-13228-des-produits-de-
chococam-filiale-du-sud-africain-tiger-brands-au-centre-d-une-polemique-au-cameroun
[24] Id
[25] Id
[26] Id

violation of the compulsory application of Standard NC 04 on the labeling of prepackaged foods in Cameroon which states that "the label affixed to prepackaged goods shall not describe or represent the product in a false, misleading, deceptive manner or likely to create in any way, an erroneous impression about its nature."[27]

According to the Foundation, the labeling of "Tartina" and "Choconut" spreads failed to indicate either the percentage of peanuts or the cocoa powder that these products contain. They did not mention of the type of vegetable oil used (e.g., palm, sunflower, or coconut) in their production, which are all necessary elements on a food label that ought to be brought to the knowledge of their consumers.

In rebuttal, *Chococam* claimed that all the ingredients used in the manufacturing of their chocolate products were listed in accordance with the standard NC 04: 2000-20 on the labeling of prepacked food products in Cameroon. They went further to lay emphasis on the fact that all of their chocolate spreads, including all of the food products marketed by *Chococam*, as well as their labeling, were subjected to conformity assessment by ANOR. However, regarding the mentioning of the label "the brand of chocolate spread" on their website, they did take note of that point and undertook to make the necessary modifications.

Source: https://www.kimoliamarket.com/product/tartina-melted-chocolate-28kg/

---

[27] Alphonse Ayissi Abena, president of Focaco(Fondation camerounaise des consommateurs) citing ANOR Standard NC 04 https://www.investiraucameroun.com/gestion-publique/1709-13228-des-produits-de-chococam-filiale-du-sud-africain-tiger-brands-au-centre-d-une-polemique-au-cameroun. See also L'Étiquetage des Denrées Alimentaires Préemballées (CODEX STAN 1-1985), Article 3.1. "L'étiquette apposée sur les denrées préemballées ne devra pas décrire ou présenter le produit de façon fausse, trompeuse, mensongère ou susceptible de créer d'une façon quelconque une impression erronée au sujet de sa nature veritable." Available at http://www.fao.org/3/a1390f/a1390f00.pdf (Last accessed 2/14/2020).

Incidences such as these show that consumer associations in Cameroon have been at the forefront in the fight to endure that the food products produced in Cameroon for consumption respectfully adhere to all the standards of labeling enforceable by the law.

E. Consumed foods of animal origin

The diet in Cameroon mostly consists of foods of animal origin. These include a wide variety of dairy, beef and fish products, among others. There is equally high consumption of honey generally used to replace sugar and in the formulation of traditional medicine. Although these products are highly valued for their rich nutritive content, they are mostly noted for their potential to carry contaminants. Production conditions and application of improper procedures during processing greatly affect the quality and safety of these products. This calls for more stringent monitoring protocols for the analysis of antibiotics residues, sulfonamides, pesticides, and heavy metals to meet standards. For example, it is common practice to dilute pure honey with a little amount of water and sometimes caramelized sugar before selling to boost quantity and profit. This adversely affects the quality of the honey and its shelf life.

Increasing consumer awareness of food risks and trade globalization are boosting the production of foods of animal origin such as milk and honey. However, most consumers struggle to interpret quantitative information contained in labels of milk products. Meanwhile, some find different nutrition label formats confusing, as well as too much information provided on the label of foods of animal origin overwhelming. Yet, some consumers have also indicated that they regard the taste of a product as being more important than its nutritional content. This emphasizes the need to educate consumers on how to make healthier food choices, while utilizing the information provided on the food label, but within the boundaries of the aforementioned factors.

In recent years, the global trend has been to impose mandatory nutrition labeling on all products, regardless of whether they pose a health or nutrition hazard or not. To reflect this trend, the Codex Alimentarius guidelines has been adopted in Cameroon in addition to local legislature to recommend that nutrition labeling should be mandatory, even in the absence of health claims.

F. Organic and natural products

Another growing concern necessitating stricter regulation on the labeling of food products for mass consumption is in the domain of homemade organic and natural fruit juice and dairy products. Due to the health consciousness gaining grounds all over the world today, people are now are opting for the consumption of more natural and organic products. As a result, many of today's consumers are demanding greater transparency in their food products and purchases. Consumers seek to know where their products come from and what exactly they are made of. Consequently, with this lifestyle shift, the impact on the food industry is obvious, as market leaders scramble to offer genetically modified organism (GMO)- and additive-free offerings.[28]

In the United States alone, according to the USDA,[29] consumer demand for organically produced goods continues to show double-digit growth. In fact, in 2018, the U.S. organic products market broke the record by exceeding the $50 billion mark for the first time, according to the Organic Trade Association.[30] One might wonder if these trends are also reflected in Africa and the answer is a resounding yes, they are. For example, although relatively small, the South African organic foods market is reported to be steadily growing as evidenced by the increase in certified farms from 35 in 1999 to 250 in 2018.[31]

Likewise, the organic trend is gaining a stronghold in Cameroon although more so in a bid to alleviate poverty and reduce unemployment. The government has facilitated the creation of empowerment centers for women and young entrepreneurs that want to engage in small startups in the food industry. One of these government-run support programs is known as Agropole (Agricultural growth poles), which provides not only training but also funds for farmers to process and market, for instance, spices and other forest-grown plants, to boost incomes and jobs, while conserving trees and limiting climate change.[32] These farmers in turn return to their communities to provide informal education to other farmers to spread the knowledge. Though these private initiatives are laudable, they however carry with them certain inherent risks with respect to the hygiene standards and norms required for the commercialization of these products.

According to a 2017 report by the International Institute for Sustainable Development, Africa has seen the emergence of thirty-six agricultural growth poles and nine corridors over the past fifteen years.[33] As of May

[28] https://www.investopedia.com/articles/investing/022217/study-shows-surge-demand-natural-products.asp
[29] https://www.ers.usda.gov/topics/natural-resources-environment/organic-agriculture/organic-market-overview/
[30] https://ota.com/news/press-releases/20699
[31] https://www.fas.usda.gov/data/south-africa-growing-trade-opportunities-us-and-south-african-organic-food
[32] https://www.weforum.org/agenda/2019/04/cameroon-farmers-spice-up-earnings-with-forest-friendly-foods/
[33] https://www.iisd.org/sites/default/files/publications/rise-agricultural-growth-poles-in-africa.pdf

2017, Cameroon alone had established 40 small projects although with mixed results, mainly due to a strong focus on increasing productions without the corresponding investments in improving market access and distribution channels.[34] While the organic sector in Cameroon is still extremely underdeveloped, estimates of certified organic production suggest that 7,000 hectares of land are under organic management accounting for 0.08 percent of the total agricultural area.[35] Cameroon therefore shows a strong potential for further developing the organic sector. What is even more interesting is the fact that with high premiums for organic produce, some areas in Cameroon have spontaneously converted to organic farming entirely.[36]

As consumption and demand for organic products increase, so too is the need for processing through drying, pulping, and juicing. Generally, these prepackaged homemade natural juices and nonalcoholic beverages, almost never carry any form of labeling. Their packaging is comprised mostly of used plastic carbonated drink or purified water bottles. This is as alarming as their noncompliant production, which is not subject to any mechanism of control or recycling before usage. These products are often sold by hawkers on the streets thereby proliferating public consumption without any form of certification by ANOR. Besides regulation by ANOR, Ecocert IMOswiss AG[37] also provides certification services according to organic, environmental, and social standards in Cameroon.[38] Meanwhile, Association for the Promotion of Organic Agriculture in Cameroon (ASPABIC) in the French-speaking regions and Association of Vegetable Growers (AVEGRO) in the English-speaking regions support the promotion of the organic sector in Cameroon by raising public awareness. They provide their members with information services, technical assistance, and training. However, the labeling concerns remain problematic.

One common example of such widely sold homemade organic juices is locally known as *folere*.[39] It is a drink made from hibiscus petals very common in Africa and known by different names in different countries: *zobo* in Nigeria,[40] *sobolo*[41] in Ghana, *bissap* in Senegal,[42] *karkade* in North Africa (Egypt and Sudan). It is also consumed in the Caribbean, where

---

[34] Id
[35] Id
[36] Id
[37] https://www.ecocert-imo.ch/logicio/pmws/indexDOM.php?client_id=imo&page_id=home
[38] http://www.intracen.org/exporters/organic-products/country-focus/Country-Profile-Cameroon/
[39] https://afrovitalityeats.com/recipe/folere-cameroon-hibiscus-ice-tea/
[40] Id
[41] https://www.tasteatlas.com/sobolo
[42] https://afrovitalityeats.com/recipe/folere-cameroon-hibiscus-ice-tea/

it is commonly known as *sorrel*.[43] Folere is a local favorite especially during hot summer days, and because of its purported medicinal value in
reducing high blood pressure.[44] However, folere is often prepared with
tremendous quantities of sugar which usually far exceed the daily recommended caloric intake. The local manufacturer can create their recipes to
make their folere as sweet as they like.

Folere is usually sold in plastic bottles, which carry no labeling of the
content or nutritional information of the production process are filled to
the brim and seating in ice buckets. The consumer can pay a little extra
to keep the bottle but oftentimes customers will gulp down a bottle right
next to the stand. Without any labels specifying the sugar content, people
tend to consume an otherwise "organic" product, which is presumably
healthy, but which could actually be extremely unhealthy due to its high
sugar content. For individuals with health conditions such as diabetes, this
could lead to diabetic comas and other sugar-related health complications.

Source: https://www.tasteatlas.com/sobolo

Moving forward: Control mechanisms
In 2016, the Cameroon Food Industry Technical Centre (Centre Technique
de l'Agro-alimentaire du Cameroun, CTA-CAM) was created as an analysis center for processed food in Cameroon.[45] It will mainly oversee the
application and respect of the standards established in the country in
terms of food industry, before the different products are available on the

[43] http://www.jamaicatravelandculture.com/food_and_drink/sorrel_drink.htm
[44] https://www.medicalnewstoday.com/articles/318120
[45] https://www.journalducameroun.com/le-cameroun-dispose-dun-centre-technique-de-
lagro-alimentaire/

market. Managed by the private sector, CTA-CAM is planning to improve the competitiveness of the Cameroonian companies and guarantee food safety in the country. Although it may be too soon to assess the impact of this center's role in upholding food labeling and safety, strides such as these seem to suggest that when it comes to labeling and food safety, Cameroon seems to be headed in the right direction.

Likewise, in 2017, ANOR established the annual work plan for the development of national Cameroonian food labeling standards.[46] This work plan, related to the process of developing national standardization programs, instituted by Circular No. 008/PM of November 12, 2010,[47] aims to build an efficient standardization system in order to support government policy and increase competitiveness of local businesses, to guarantee the health and safety of consumers and to preserve the environment. This program is expected to contribute to the rapid development of standardization in Cameroon (substantial increase in the number of relevant standards, updated and accessible to interested parties) with a view to catching up with countries with comparable levels of development.

For effective food labeling to be realized in Cameroon, enormous sensitization and education campaigns need to be implemented to educate and change consumer and producer habits to foster compliance with labeling standards and improve product quality assurance, safety, and reliability. Consumers and producers alike must be able to comprehend that complying with food labeling standards is important to not only guarantee that consumers make informed choices on the food they buy and consume but that labels help reduce health and safety concerns and improve quality of life in general.

[46] http://www.anorcameroun.info/index.php/article/view/national-standards-development-program-the-annual-work-plan-is-validated-by
[47] Id

Index

Note: Locators in *italics* represent figures and **bold** indicate tables in the text.